Environmental Learning

Mark Rickinson • Cecilia Lundholm
Nick Hopwood

Environmental Learning

Insights from research into the student experience

Springer

Mark Rickinson
Educational Research Consultant
Visiting Research Fellow
Oxford University and Policy Studies Institute
The Innovation Centre, Howbery Park
Wallingford, OX10 8BA, UK
m.rickinson@dsl.pipex.com

Cecilia Lundholm
Research Fellow
Department of Education and Stockholm Resilience Centre
Stockholm University
SE-106 91 Stockholm
Sweden
cecilia.lundholm@ped.su.se

Nick Hopwood
Research and Evaluation Officer
Centre for Excellence in Preparing for Academic Practice
Oxford Learning Institute
Littlegate House
16/17 St Ebbe's Street
Oxford, OX1 1PT, UK
nick.hopwood@learning.ox.ac.uk

ISBN 978-90-481-2955-3 e-ISBN 978-90-481-2956-0
DOI 10.1007/978-90-481-2956-0
Springer Dordrecht Heidelberg London New York

Library of Congress Control Number: 2009927908

Printed on acid-free paper

Springer is part of Springer Science+Business Media (www.springer.com)

This book is dedicated to Graham Corney (1943–2008).

Contents

Introduction

This book is about how school and university students experience and respond to learning activities concerned with environmental issues. While the learning demands associated with sustainable development become ever greater and more complex, our understandings of the nature and dynamics of such learning are in their early infancy. Environmental education and education for sustainable development have become features of many countries' formal education systems, but very little is known about what such provision looks and feels like for the learners concerned. The aim of this book is to bring learners and their experiences to the centre of current debates about environmental education and education for sustainable development. By exploring the real-time actions, interactions and interpretations of individual learners in various environmental learning situations, we show how insights from research into the student experience can provide powerful pointers for future practice, policy and research.

Education and the Environment

The last 4 decades have seen growing international recognition for the educational dimensions of environmental and sustainable development issues. Since the late 1960s, international statements from organisations such as the IUCN (International Union for the Conservation of Nature) and UNESCO (United Nations Educational, Scientific and Cultural Organisation) have called for environmental problems to be tackled through environmental education for all age groups. The 1972 UN Conference on the Human Environment in Stockholm declared that: 'education in environmental matters, for the younger generation as well as adults [...] is essential' (United Nation 1972).

The nature and goals of such environmental education were given further shape by subsequent international events in Belgrade and Tbilisi. The Belgrade Charter (UNESCO 1975) stated that:

> The goal of environmental education is to develop a world population that is aware of, and concerned about, the environment and its associated problems, and which has the knowledge,

M. Rickinson et al., *Environmental Learning: Insights from Research into the Student Experience*, DOI 10.1007/978-90-481-2956-0_1,

> skills, attitudes, motivations, and commitment to work individually and collectively toward solutions of current problems and the prevention of new ones.

Meanwhile, the Tbilisi Declaration (UNESCO 1978) reportedly 'established a framework for an international consensus which without doubt has been the seminal influence on the development of environmental education policies around the globe' (Palmer 1998, p. 8).

Subsequently the 1987 World Conference on Environment and Development and the 1992 United Nations Conference on Environment and Development further underlined the potential role of education in relation to sustainable development. The main outcome of the latter was very clear that 'Education is critical for promoting sustainable development and improving the capacity of the people to address environment and development issues' (UNCED 1992). More recently still, the United Nations declared 2005–2014 the Decade for Education for Sustainable Development with the goal of 'integrating the principles, values, and practices of sustainable development into all aspects of education and learning' (UNESCO 2005, p. 5).

So over the last forty years environmental education, and more recently sustainable development education, have grown in prominence across a range of different formal, informal and non-formal educational contexts. These concepts are now present in international education frameworks, national education policies, national curriculum documents, higher education strategic initiatives, school curriculum developments and international conservation strategies. As reported by the Fourth International Conference on Environmental Education:

> Many thousands of environmental education programmes, projects and materials have been developed and used with millions of learners, community members and decision makers in all countries and learning contexts around the world. (Fourth International Conference on Environmental Education 2007, p. 3)

What this and other similar recent conferences (e.g. UNESCO 2007; OECD 2008) also make clear, though, is how the international economic, social and ecological challenges that necessitate such efforts have worsened not improved over recent years. In other words, against the backdrop of international financial crises, poverty and social justice issues and global climate change, the need for widespread, effective environmental education and education for sustainable development has never been greater. Within this broad setting, our focus is on environmental learning. 'Environmental learning' as a term has varied academic meanings and presumably multiple vernacular interpretations. It can refer to learning in or through particular (often outdoor) environments, environmental cognition (e.g. Kaplan 1985), as well as pointing to a broad conceptualisation of learning that occurs when learners engage with content that pertains in some way to the environment (not just natural or physical) or environmental issues. The latter frames the approach adopted in this book, although we focus specifically on formal learning settings, and the chapter titled "What Is Environmental Learning?" explores our perspective in more detail.

The Role of Learning

Perhaps surprisingly, given the developments noted above, there has not been a great deal of specific discussion about learning in relation to environmental and sustainable development education. Until recently, questions concerning the forms of learning that are needed or that might be taking place in environmental and sustainable development education have rarely been raised in any sort of explicit way. This situation is changing, however, as certain writers have sought to place learning at the centre of discussions about the environment and sustainability. Perhaps the clearest example of this is Scott and Gough's (2003) volume on *Sustainable Development and Learning* which works from the basis that:

> It is not enough to say that sustainable development and learning need to go hand in hand; rather sustainable development itself needs to be understood as a learning process (p. xiv).

Sterling (2001) similarly is clear that 'the difference between a sustainable or a chaotic future is learning' (p. 10). The titles of several other recent publications also suggest a developing emphasis on learning: *Social Learning Towards a Sustainable World* (Wals 2007); *Participation and Learning: Perspectives on education and the environment, health and sustainability* (Reid et al. 2008); *Engaging Environmental Education: Learning, culture and agency* (Stevenson and Dillon forthcoming).

But why is a growing emphasis on learning significant? Scott and Gough (2003) highlight three reasons why learning is an important focus for attention in environmental and sustainability education. First, unlike education or training, learning is not restricted to any particular stage of life or institutional setting, that is, it is both life-long and life-wide. Secondly, learning is important if one believes that sustainable development is inherently about learning. Scott and Gough argue that 'there will be no sustainable development where learning is not happening' (p. xiv). Thirdly, learning is about 'what learners learn [as distinct from] what teachers teach' (p. 38). In other words, it includes the 'serendipitous and accidental' and the 'negative as well as positive' (p. 38).

We would add that a focus on environmental learning is important because it is a topic that has been neglected by educational researchers. During the 1990s, for example, there were many studies investigating characteristics of school students (e.g. what kinds of environmental attitudes or knowledge they have) but few exploring the process or outcomes of environmental learning (Rickinson 2001). There has also been neglect in a more subtle sense in that, where learning has been studied or discussed, it has tended to be in relation to educational interventions rather than as a process in its own right (Rickinson 2006). In other words, concern with environmental learning as 'an integral part of our everyday lives' (Wenger 1998, p. 8) and with 'what learners learn, not with what teachers teach' (Scott and Gough 2003, p. 38) has been all too rare. There have also been few attempts to 'develop new models of [environmental] learning' due to an apparent reluctance by environmental education researchers 'to engage with learning theories' (Dillon 2003, p. 217).

The key consequence of all of this is that we know very little about the nature and dynamics of learning in environmental and sustainability education and, most importantly, how such learning is experienced by the learner.

The Student Experience

This book is primarily concerned with how students in schools and universities make sense of and respond to learning focused on environmental issues. It is about what environmental learning is like from the perspective of those who are doing it. As such, it can be seen as a response to significant gaps within the environmental education research literature. As highlighted by Payne some time ago, there has been 'a lack of consideration in environmental education theory and research practices about the children who are the subjects of environmental education' (1998, p. 20). While there has been some progress on this front in terms of the emergence of new research on environmental learning (see the chapter titled "Researching Environmental Learning"), there is still much more written about what environmental education *ought* to be like as opposed to about what environmental education *is* actually like in everyday learning contexts. There also continues to be 'a marked predominance of evidence on learning *outcomes*, but very little about learning *processes*' (Rickinson 2001, p. 216). Much has been said about the impact of educational programmes in terms of pre-determined learning outcomes, but little about learners' experiences of and responses to these programmes.

The research reported in this book draws on the distinction between the curriculum as laid out in policy documents (the *specified* curriculum) and as operationalised by teachers (the *enacted* curriculum) and the curriculum as encountered by students (the *experienced* curriculum) (Goodlad and associates 1979; Banks and McCormick 2005). Our focus is very much on the third of these, that is 'the subjective experience of students as they are engaged in learning' (Erickson and Shultz 1992, p. 446). More specifically, we are interested in students':

- Affective responses to the curriculum – Watts and Alsop (1997, p. 355–356) stress the importance of considering 'not only what conceptual systems learners hold and the status that can be attached to them, but also *how they feel about this knowledge* as well' (emphasis added).
- Interpretations of learning tasks – As Nespor (1985 cited in Doyle 1986, p. 366) has argued, 'There is the task as the teacher has it in his or her mind, the task as the teacher announces it initially to the students, the task as it is eventually negotiated in teacher–student interactions and *the task as each student comes to understand it in light of his or her background and interests and expectation*' (emphasis added).
- Learning strategies and processes – Cooper and McIntyre's (1996, p. 159) work on students' craft knowledge indicates that students 'are a vital source of information about their own learning processes and the ways in which contextual factors (classroom, task, peer and teacher variables) interact with these processes'.

In the context of environmental and sustainability education, this book is about uncovering 'what the student is up to [...] in daily classroom encounters with the curriculum' (Erickson and Shultz 1992, p. 468). We see this as a productive response not only to the acute shortage of detailed empirical work on students' environmental learning but also to more general calls for 'more studies which examine the particular discipline-based characteristics of the student learning experience' (Ertl et al. 2007, p. 69). It also fits well with UNESCO's (2005, p.20) recommendation for 'increased research on quality teaching and learning approaches' as part of the UN Decade of Education for Sustainable Development.

The Evidence Underpinning This Book

This book is based on in-depth empirical research with students in school and university classrooms. In particular, it draws on studies undertaken by ourselves in the United Kingdom and Sweden. The findings from our work, however, are discussed in connection with wider international research on student learning both within and beyond environmental education.

Our own studies were undertaken independently but drew on similar theoretical perspectives and methodological approaches. The focus of Rickinson's (1999a, b; Rickinson and Robinson 1999) work was on the ways in which 3 teachers and 12 students (age 13–15) in 3 UK secondary schools dealt with controversial environmental issues within geography. It was a qualitative investigation into: (i) the ways in which environmental curriculum topics were being taught by the teachers, and the thinking that lay behind these practices; (ii) the ways in which such teaching was experienced by their students; and (iii) the similarities and/or differences between the perspectives of the teachers and the students. These foci were explored in the context of lessons on environmental topics that were perceived by the teachers to be issues-based (as opposed to purely factual). Examples included lessons on rainforest development, nuclear power and indigenous peoples. Data were generated through lesson observation and audio-recording, student lesson impression sheets and post-lesson interviews with students and teachers. Central to these methods was a desire to understand the students' and teachers' actions from their perspectives.

Lundholm's (2003, 2004a, b, 2005, 2007, 2008a, b) research examined Swedish university students' learning about environmental issues as part of undergraduate and graduate programmes in civil engineering, biology and economics. It consisted of case studies based on different kinds of environmental content: (i) first-year engineering students' interpretations and learning during a compulsory ecology course; (ii) biology students' learning during an environmental auditing task, which encompassed issues relating to economics, business and administration; (iii) doctoral students' interpretations of the task of writing a thesis in the environmental field, focussing on natural science, social science or both; and (iv) biology graduate students' learning experiences in economics. Through observation and in-depth interviewing, this work sought to explore students' learning processes and the ways in which they interpreted and experienced tasks and courses in the environmental field. The transcripts of interviews and group discussions were analysed from an intentional perspective, that is, a perspective which takes into account the students' aims defined as 'projects' (Halldén 1988; Halldén et al. 2007).

Finally, Hopwood (2006, 2007a, b, c, 2008, forthcoming) worked in three UK secondary schools following the experiences of six students in their geography lessons for a period of 3 months in each school. The students – Bart, Lisa, Matt, Sara, Ryan, and Jenie – were in year 9 at the time of research (age 13–14 years). The study combined a mixture of ethnographic techniques with a multiple case-study approach, in which each of the students was treated as a separate case. A suite of methods produced data in a range of forms: loosely structured field notes; copies of handouts, worksheets etc.; transcripts of one-to-one interviews with students conducted after each lesson; and transcripts of researcher–student interactions structured around specifically designed activities (including concept mapping and self-directed photography; see Hopwood 2006). In total nearly 90 lessons were observed, and over 70 interviews conducted, at least 11 per case student.

Taken together, these three studies span a number of educational contexts (English secondary school and Swedish university), curricular subjects (geography, civil engineering, biology and ecology), learner types (13–15 year old school students and 20–45 years university students) and environmental content (for example, business and the environment, climate change, rainforest development, migration, waste management, etc.) (see the chapter titled "What Is Environmental Learning?" and Appendix I: Empirical Context for more details). Although originally undertaken independently, all three share a number of important commonalities in terms of their:

- Focus on the student perspective – All had a clear focus on accessing experiential accounts of students' environmental learning experiences, underpinned by a view of students as active agents rather than passive recipients.
- Conceptualisation of curriculum processes – All drew on the distinction between task as presented by the teacher (the enacted curriculum) and the problem (or 'project') as understood by an individual learner (the experienced curriculum) (Halldén 1988; Erickson and Shultz 1992; Entwistle and Smith 2002).
- Methodological perspectives – All used similar case-study approaches and qualitative techniques including observation/recording of classroom talk, documentary analysis of students' work and interviews/focus groups about particular learning experiences.

While these three studies provide the main empirical underpinning of this book, the central arguments are significantly strengthened by insights from wider research within and beyond the field of environmental learning. We draw, for example, on other studies in environmental education that have examined students' perspectives on geography fieldtrips (Lai 1999), nature-based excursions (Ballantyne and Packer 2002) and primary school experiences (e.g. Nagel and Lidstone 2008). We also make connections with studies of environmental learning in non-school settings such as outdoor centres and national parks (e.g. Brody and Tomkiewicz 2002), as well as conceptual papers that have examined different perspectives on learning in the context of environmental education (e.g. Dillon 2003; Falk 2005).

Equally important, though, are ideas and findings from beyond the field of environmental education. We make links with classroom researchers who have examined students' craft knowledge (e.g. Cooper and McIntyre 1996), curriculum researchers who have asked about students' experiences (e.g. Lord and Jones 2006), science education researchers who have investigated students' affective responses to socio-scientific issues (e.g. Watts and Alsop 1997), teacher education researchers who have looked at how teachers help students to make emotional connections to subject matter (e.g. Rosiek 2003) and learning researchers who have asked questions about the role of emotions and feelings (e.g. Efklides and Volet 2005). In this way, we are seeking to locate research and thinking about environmental learning within the context of wider developments in research on learning, teaching, curriculum and pedagogy.

Structure of the Book

In this opening chapter, we have outlined the focus and scope of this book and explained the evidence that underpins its claims. We have also highlighted the significance of a research-based publication on students' environmental learning experiences in terms of international policy developments and gaps in current environmental education research. We hope that the subsequent chapters will allow readers to further understand the ways in which we are conceptualising and framing the topic of environmental learning (the chapters titled "What Is Environmental Learning?", "Researching Environmental Learning" and "Lenses for Understanding Environmental Learning"), to engage with the richness and detail of what we have learnt about students' environmental learning processes (the chapters titled "Dealing with Emotions and Values", "Questioning Relevance" and "Negotiating Viewpoints Among Students and Teachers") and to come away with a clear sense of what this work means for future environmental learning practice, policy and research (the chapter titled "Enhancing Environmental Learning").

The chapter titled "What Is Environmental Learning?" discusses the nature and meaning of the term 'environmental learning'. Against the backdrop of a general dearth of elaborations of this concept within the literature, we draw together the few definitions and typologies that have been helpful to us in clarifying the breadth and complexity of this topic. We then describe and explain the framework that we have used to make sense of environmental learning in formal education settings in terms of five dimensions (who, what, where, how and why). These dimensions provide a framework that we return to at the end of the chapters titled "Dealing with Emotions and Values", "Questioning Relevance" and Negotiating Viewpoints Among Students and Teachers" in order to pull out recurring findings and emerging messages.

The chapter titled "Researching Environmental Learning" considers the research landscape that surrounds and informs the work reported in this book. An important starting point for this chapter is the fact that researching environmental learning has only fairly recently emerged as a topic of concern amongst environmental education researchers. That said, three broad approaches to researching environmental learning can be identified: researching learners; measuring outcomes; and exploring processes. Our own work on students' environmental learning experiences is discussed as illustrative of the third of these categories concerned with learning processes as opposed to learning outcomes.

The chapter titled "Lenses for Understanding Environmental Learning" provides the final part of the background necessary for understanding the subsequent findings chapters. Here we introduce the idea of lenses as interpretive tools that help us to look at or for different aspects of environmental learning. We outline the three lenses that we have used to make sense of the environmental learning settings within our studies. These focus on (i) emotions and values, (ii) questions of relevance, and (iii) differing viewpoints among teachers and learners. Each of these lenses provides the basis for each of the three subsequent findings chapters.

The chapter titled "Dealing with Emotions and Values" takes the first lens and looks at students' environmental learning in terms of the role of emotions and values. We look, firstly, at students' emotions and values as part of the learning process. Examples include reactions of distaste and discomfort with particular learning activities, as well as engagement in learning sparked by learners' emotional responses towards particular issues. Secondly, we explore how students' perceptions of environmental subject matter can affect their learning. This focuses on the extent to which learners see environmental subject matter as value-laden and the ways in which they confront notions of subjectivity/opinions versus objectivity/facts.

The chapter titled "Questioning Relevance" focuses on the second of the three lenses, which looks at issues of relevance within environmental learning processes. Our analysis pursues two related themes. The first explores how environmental learning may (or may not) be conceived as relevant to learners themselves, while the second considers learners' perceptions of the relevance of environmental learning to particular curricular contexts. In both cases, learners' ideas of relevance appear to have a strong bearing on their engagement in environmental learning processes and their conceptual notions of relationships between the various things they learn.

The chapter titled "Negotiating Viewpoints Among Students and Teachers" explores environmental learning situations in terms of the interactions between the viewpoints of students and the viewpoints of teachers. Using this lens, we consider four types of teacher-student conflict situations that arose within our studies of school and university environmental learning. These concern differences in viewpoints between teachers and students on: specific environmental issues; what is topical or controversial; what is relevant for a specific curriculum subject; and the nature and value of empathy tasks. The important point here is that environmental learning activities can involve subtle, often hidden tensions between teachers and learners which can have a very real effect on the nature of students' engagement.

The chapter titled "Enhancing Environmental Learning", our concluding chapter, draws together and provides some further commentary on the key findings emerging from the chapters titled "Dealing with Emotions and Values", "Questioning Relevance" and "Negotiating Viewpoints Among Students and Teachers" and the book as a whole. It considers the implications that stem from this work for practitioners, policy-makers and researchers interested in understanding and enhancing the quality of environmental learning.

What Is Environmental Learning?

In this chapter we present a brief discussion of how the term 'environmental learning' has been defined and explained in the literature, and the complex range of activities, interests, contexts and purposes that it can encompass. We then detail what we mean and how we engage with the term, by discussing the who, what, where, how, and why of environmental learning. For each of these we illustrate how the focus of this book (which arises from the empirical contexts upon which it is based) in some ways reflects the diversity of environmental education (as with the areas of content addressed), but in others is quite narrowly focused on specific aspects (as with our emphasis on formal settings).

Defining Environmental Learning

The current literature on environmental education provides surprisingly few definitions or explanations of environmental learning. As we and others have argued (e.g. Dillon 2003; Myers 2006; Rickinson 2006; Hart 2007), the field of environmental education does not have a good record of articulating what 'learning' means in a direct manner. This is not to say that ideas about learning have not been implicit within discussions about environmental education for many years. The difficulty is that there has been little explicit discussion and writing on the topic.

An important exception to this is Scott and Gough's (2003) book *Sustainable Development and Learning*, which provides several helpful conceptual starting points for understanding the nature and scope of environmental learning. They refer to environmental learning as:

> learning which accrues from an engagement with the environment or environmental ideas (Scott and Gough 2003, p. 14).

We find the open-ness of this description of environmental learning helpful in highlighting its breadth and complexity in a number of ways. Firstly, it helps to flag up *the range of foci and outcomes* that can be associated with environmental learning. Scott and Gough (2003), for example, identify 'nine categories of interest which

M. Rickinson et al., *Environmental Learning: Insights from Research into the Student Experience*, DOI 10.1007/978-90-481-2956-0_2,

Table 1 Categories of interest in environmental learning

Categories of interest	Focus and outcomes	Exemplified by
Sharing the joy and fulfilment of nature	Nature, values and feelings	Non-formal educators and interpreters
Studying the processes of nature	Nature understanding	Teachers of ecology
Nature as a heuristic to develop knowledge/skills	Nature skills	Interpreters, field study officers
Environments as a heuristic to achieve conservation	Conservation understanding	Conservation/heritage scientists
Promoting individual behavioural change	Conservation behaviours	Teachers and environmental activists
Advocating particular modes of social change	Social change/social justice	Teachers and environmental activists
Sustainability issues as contexts for democratic change	Social change, democratic citizenship skills	Teachers and others interested in citizenship
Nature as a metaphor for a preferred social order	Social change values	Sociobiologists, deep ecologists, Gaianists
Studying environmental learning	Learning, learning about learning	Educational researchers

Source: Adapted from Scott and Gough (2003, p. 54).

capture, albeit in a rather tentative way, a range of different focuses and objectives of those who espouse and promote environmental learning'. This begins to clarify some of the foci (e.g. nature, conservation and social change) and desired outcomes (e.g. values and feelings, understanding, skills, behaviours, social justice and democratic citizenship skills) of learning associated with different kinds of environmental education (Table 1).

A second source of complexity in environmental learning is the *breadth of contexts and processes* that it can involve. As Falk (2005, p. 267) has argued, environmental learning needs to be understood in relation to a wide range of settings and information sources including 'the formal education system, books from both libraries and retail outlets, museums, parks, ecotourism sites, television programming, film and video, newspapers, radio, magazines, the Internet, community-based organisations and through conversations with friends and family'. A similar point is made by Scott and Gough (2003), who emphasise that as well as diversity of learning contexts there is also diversity of learning processes or strategies. They identify three strategies through which environmental learning can be understood to occur. 'Instruction' of learners focuses on one-way transmission of information, with agreed parameters and assumptions. 'Engagement' of learners involves communication through two-way exchanges in which the main assumptions are shared, but there may be some disagreement or contestation about others. 'Facilitation' of learning focuses on mediation through two-way or multidirectional exchanges in which important parameters and/or assumptions are disputed. Across its many possible contexts, then, environmental learning can encompass a range of different processes. What is more, as highlighted by a recent international conference report, this diversity is increasing with time:

> In 1987, the emphasis was on education and training, while discussions in 1997 introduced collaborative learning. Today [in 2007] the emphasis is on experimentation and broader social and cultural situated learning processes that take account of context (Fourth International Conference on Environmental Education 2007, p. 2).

Finally, there is complexity in environmental learning when we consider its *qualities and underlying assumptions*. Sterling (2004, p. 69) emphasises the need to 'distinguish between qualities or levels of [environmental] learning'. Using systems terms, he draws distinctions between three types of learning:

- Accommodation (first order learning, adaptation, maintenance, learning about sustainability)
- Reformation (second order learning, critically reflective adaptation, learning for sustainability)
- Transformation (third order learning, creative re-visioning, learning as sustainability) (Sterling 2001, p. 78).

His arguments in favour of 'a deeper understanding of the relationship between learning, society and sustainability' (Sterling 2004, p. 68) are helpful in highlighting the idea that environmental learning can encompass not only different foci, outcomes, contexts and processes but also different levels of depth. Alongside this, further categorisations of environmental learning are provided by Scott and Gough (2003) and Vare and Scott (2007). The former distinguish between three types of learning based on differing assumptions about the relationships between conceptions of the causes of environmental or social problems and the role of environmental learning in dealing with them (see Box 1). Vare and Scott (2007), meanwhile,

Box 1 Learning described as Type 1, 2 or 3

Type 1 approaches assume that the problems humanity faces are essentially environmental, and can be understood through science and be resolved by appropriate environmental and/or social actions and technologies. It is assumed that learning leads to change once facts have been established and communicated.

Type 2 approaches assume that our fundamental problems are social and/or political, and that these problems produce environmental symptoms. Such fundamental problems can be understood by means of anything from social-scientific analysis to an appeal to indigenous knowledge. The solution in each case is to bring about social change, where learning is a tool to facilitate choice between alternative futures which can be specified on the basis of what is known in the present.

Type 3 approaches assume that what is (and can be) known in the present is not adequate; desired 'end-sates' cannot be specified. This means that any learning must be open-ended. Type 3 approaches are essential if the uncertainties and complexities inherent in how we live now are to lead to reflective social learning about how we might live in the future.
(Scott and Gough 2003, p. 111–116).

identify two complementary approaches to 'learning for a change' in education for sustainable development. The first (ESD 1) promotes learning that focuses on informed and skilled behaviours and ways of thinking in circumstances where needs are clear and agreed. The second (ESD 2) is described as 'building capacity to think critically about what experts say and to test ideas, exploring the dilemmas and contradictions inherent in sustainable living' (Vare and Scott 2007, p. 191). The authors stress that both the approaches are necessary and are not incompatible. On the one hand, without knowing the facts, one cannot cast value judgements or critically analyse what measures are appropriate to take at a specific time and place. On the other hand, the 'uncertainty' of facts in light of a future that is largely unknown, highlights the need for critical thinking and open-ended learning.

Taken together these various categorisations of learning in the context of environmental and sustainable development education serve to further underline the complexity involved in defining environmental learning. Our aim in this opening section has been to highlight the breadth of environmental learning in terms of foci, outcomes, contexts, processes, qualities and underlying assumptions. This is helpful in introducing some of the key tenets of our own understanding of the term: that it is multifaceted, not reducible to a single 'signature pedagogy', and often involves confronting issues or questions which are controversial or about which there may be few agreed-upon starting points. We now consider the framework that we have used to understand and work with environmental learning, and situate this framework in relation to wider discussions in the literature.

Our Framework

In our research we have explored the process of environmental learning by asking questions such as: How do students view particular aspects of subject matter? How do they feel about that subject matter, particular tasks, and their learning more generally? What do they do in classrooms, and why do they do these things? Our interest in understanding students' experiences is rooted in social constructivism in that we view learning as a process and product of the individual interacting with others in a social setting. In our case that setting is also a formal (institutional) one comprising peers and teachers. Having stressed the social aspect, our interest is still in the ways the individual student comes to engage, disengage and react to social and institutional norms, as for example in feeling obliged to have an opinion, or not feeling comfortable in expressing one. We view learning as a process of change in the way we look upon the world – our thoughts, feelings and actions. We see it as a process that is dependent on the learner, the object of learning and the physical/ecological, social, cultural and economic situation and setting.

In this section we present different dimensions of environmental learning that we believe are important to bring in to focus in order to come to a better understanding of students' learning. We have considered various dimensions in terms of who, what, where, how and why of learning, as represented in Fig. 1. These have

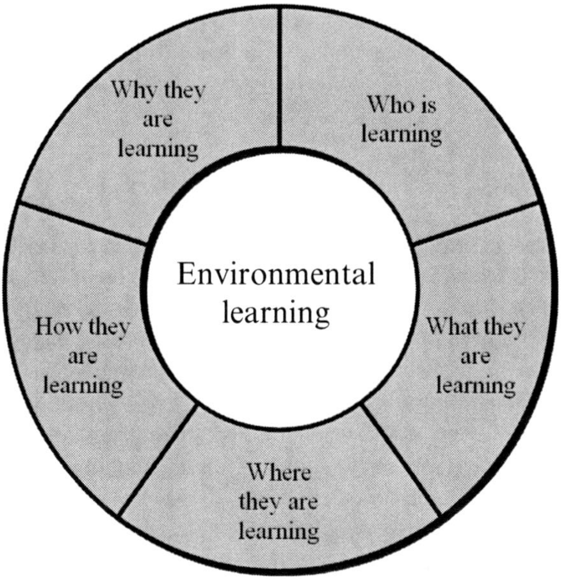

Fig. 1 The environmental learning wheel

not formed separate and distinct research questions in our inquiries, but rather we have remained conscious of them throughout our fieldwork and analysis. In particular they remind us that environmental learning is defined and shaped by much more than content (what it is about). Scott and Gough (2003; as cited at the beginning of this chapter) offer a succinct way of pointing to the scope of environmental learning: their statement remains deliberately open, avoiding narrow suggestions of content, context or method. We offer Fig. 1 as our way of being explicit about the various aspects we feel are important to think about. It also serves as a useful vehicle for us to reflect on how the empirical contexts upon which this book draws relate to environmental learning more generally.

Reflecting on the who, what, where, how and why of environmental learning, we recognise well-documented variation: it includes people of very different ages, such as children in nurseries taking part in outdoor excursions, elderly people trying to understand climate change as represented to them through television or radio, and farmers changing their production scheme and use of pesticides in favour of biodiversity. In this book, however, we narrow our focus to students' learning in formal settings. The following paragraphs take each segment of Fig. 1 in turn, highlighting relevant aspects of our empirical work, recognising that our empirical base represents in many ways only a small part of a much larger, more diverse whole. In this way we summarise key aspects of our studies and identify how our focus fits into the broader notion of environmental learning. We do not present the idea of attending to the who, what, where, how, and why of environmental learning as totally novel (others who have referred to some or all of these terms include Tilbury 1998; Scott and Gough 2003; Nagel 2004; Nagel and Lidstone 2008; Alexander et al. forthcoming).

Further information about the samples and classroom settings of the various studies is provided in Appendix I: Empirical Context.

Who Is Learning

Learners are at the centre of our conception of learning, as active participants in a process which may often be highly individualised. Learners bring their past experiences, their values, opinions and interests, and their conceptions of subject matter and curricular contexts with them when they engage in learning (although these may be in a state of perpetual flux). Learners also interpret learning experiences, tasks and activities, and perceive varying value in them. As we suggested above, the 'who' of environmental learning may include almost anyone – very young children, school and university students, adults, and the elderly.

Throughout the chapters titled "Dealing with Emotions and Values", "Questioning Relevance" and "Negotiating Viewpoints Among Students and Teachers" we highlight different ways in which environmental learning is experienced differently by different learners, within and across settings. Our research incorporated demographic variations. The UK-based studies focused on 13–14 year olds, while the Swedish research explored the learning experiences of undergraduate (20–25) and postgraduate (25–45) students. The samples in all studies included learners of both sexes.

Where They Are Learning

Learning does not take place in a vacuum. It is always located in some kind of context (social, cultural, economic, physical, etc.). Notions of context draw attention to varying degrees of formality (as explored in *Environmental Education Research* 2005 11(3)) and also indoor/outdoor settings (e.g. Rickinson et al. 2004). Our focus in this book is specifically on environmental learning in formal (classroom) contexts. However within this, we suggest that questions of 'where' also relate to the disciplinary or subject setting, what we term curricular context.

As discussed previously, environmental learning cuts across many different learning contexts and settings, and within formal education is not located entirely in one subject or discipline. At the school level it is relatively uncommon for environmental education to be experienced by students as a discrete subject. School subjects such as science and geography often provide the main curricular contexts in which environmental learning takes place. For example in England the Geography National Curriculum (DfEE/QCA 1999; QCA 2007) makes it clear that geography is a focus within the curriculum for learning about issues to do with the environment and sustainable development. Huckle (1994) argues that geography is assigned with the major responsibility for these issues in the UK curricular framework, but not exclusively so. Connell et al. (1999) found biology and geography to be the primary contexts in which young Australians said they learned about the environment, and in the United States most respondents in a large study nominated science as the main locus of their environmental learning (Roper Starch Worldwide 1994). Although at the university level some entire degrees are explicitly environmental in focus, it is common for students to undertake environmental learning as part of another degree or programme.

Such variation in curricular context is reflected in the research upon which this book is based. We investigated environmental learning in the context of geography (at the secondary school level), ecology courses (in the context of a degree in engineering) and biology (again at university level). These different contexts also incorporate variation in that the former is compulsory at the age-levels we investigated and non-vocational, while the latter two were part of an optional, vocational pursuit chosen by the learners.

What They Are Learning

The scope of environmental learning as discussed earlier can encompass wide-ranging subject matter, and there is considerable current interest in the nature of what is learned in environmental education. Environmental learning may encompass a hugely diverse content, from the form and function of physical systems (such as ecosystems or hurricanes) to global-scale interactions between people and the physical environment (such as global climate change). Environmental learning is thus not reducible to accumulation of facts or transmission of a fixed body of knowledge: it often implies dealing with the values and perspectives of oneself and others.

Boyes and Stanisstreet (1996) suggest that many of the issues confronted in environmental learning present challenges because they are invisible, intangible, uncertain, slow-moving, and multidisciplinary. They often defy straightforward scientific explanations and 'right' answers (Morgan and Lambert 2005). We would thus describe much but not necessarily all of the 'what' of environmental learning as complex, contested, and controversial. Complexity suggests a range of scales, multiple and non-linear relationships, and presents challenges for prediction as in the case of global climate change. We also note that environment-related content also poses challenges relating to defining and understanding particular problems, processes or relationships, identifying their causes, and thinking about solutions, actions, or responses. These different aspects may require different modes of thought or inquiry perhaps rooted in different disciplines or transcending them. Environmental subject matter does not always fit neatly into curricular or disciplinary structures, a situation which is not without its consequences. Österlind (2005) showed how students enrolled in explicitly multidisciplinary environmental courses encountered difficulties in negotiating the 'what' of environmental learning, questioning how to 'frame' concepts in terms of subjects or disciplinary ideas they were familiar with.

Our book explores learners' experiences in relation to a wide range of environment-related subject matter. While this inevitably does not exhaustively cover what might be learned in relation to the environment, it does incorporate sufficient diversity to reflect the overall nature of the field. Across our studies, we explored learning about: business and the environment; climate change; countries, regions and environments; economics, price and value; energy supply; environmental management and auditing; global trade and fair trade; human–nature relations; landscape; natural hazards; physical systems, processes and cycles; population structure, change and migration; rainforests; risk management; sustainable development; waste management; and weather and climate (see Appendix I: Empirical Context for more details).

Given this breadth, we find it helpful to conceptualise the 'what' of environmental learning in terms of a model which is framed around three aspects and the relationships between them, as shown in Fig. 2. In this model we are not proposing a grand theory, nor that environmental education can be reduced to a number of separate components, nor do we imply rigid or watertight divisions between the different aspects. Its value to us is in conveying the broad scope of what we have in mind when we think about the 'what' of environmental learning. 'Nature' and 'society' highlight that we think not only of physical or natural aspects, but human, cultural, political and economics aspects too (and the relationship between these). We also believe it is important to recognise that in the context of environmental learning, learners often learn much about themselves – yet individual learners are often overlooked in discussions of 'what' environmental learning is about.

The individual aspect (1) points to learning about oneself – one's beliefs, attitudes and ideas. The second aspect points to conceptualisation of and learning about nature (2), while the third highlights learning about social, economic and political structures, their functions, institutions, and ideas of justice and democracy (3). In 4, 5, and 6, the model highlights the relationships between these. The relation between individuals and nature (4) encompasses learning about how individuals affect nature, and how nature affects them. The relation between nature

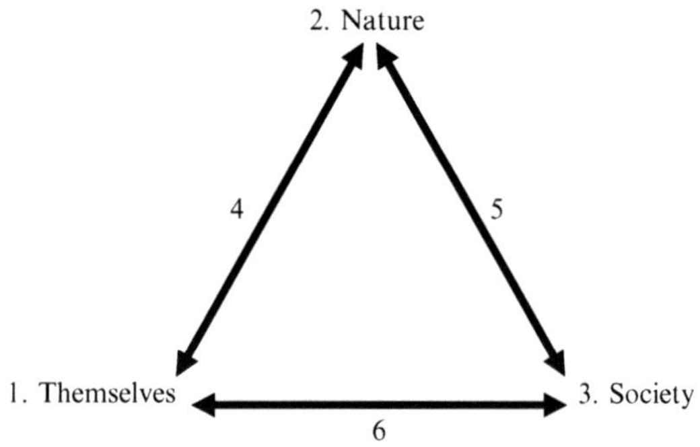

Fig. 2 Conceptualising what learners learn about in environmental learning (adapted from Lundholm 2008a)

and society (5) again explores influences in both directions, as when learners investigate the relationships between cultural practices (such as energy consumption patterns) and nature. This model finally suggests that environmental learning also involves learning about relations between individuals and society (6), how individuals situate themselves, their values and their behaviours in relation to a wider social context. Again this is bi-directional, exploring how individuals are influenced by societal norms, but also how individuals play a role in shaping and reshaping society. The learning experiences (as described by learners) which formed the focus of our empirical work spanned all six aspects of this model, although single experiences were not always exclusively associated with one aspect. This is consistent with our view of the model as an heuristic tool rather than a direct depiction of reality.

How They Are Learning

This question highlights the fact that even within the confines of formal educational settings, a range of pedagogic approaches may be evident, as documented in the literature. Our focus here is not on cognitive functioning, but rather on how learning activity is supported by teachers who set up particular experiences, tasks and activities. The focus is very much on the lived, practical experience of 'doing' in the classroom. The students in our research engaged in learning through a rich variety of tasks and experiences. These included lectures, extended group work, role-plays, debates, empathy-based writing, report writing, diagram construction, reading comprehension, poster design, poetry, and numerous work-sheet based exercises. The purpose here is to draw attention to our view that this aspect – the 'how' of environmental learning – must be attended to in our interpretation of the learning experience from the perspective of the learner, and that our empirical base for exploring these issues is indeed rich, albeit far from exhaustive.

Why They Are Learning

We have already discussed differing conceptions of environmental learning that have different purposes or outcomes in mind. Scott and Gough (2003) present a three-way typology of learning purposes and associated outcomes, and Vare and Scott (2007) distinguish different purposes with specific reference to education for sustainable development. Other writers have distinguished between education with vocational purposes from more general fostering of knowledge and understanding (e.g. Schnack 2008). It has also been suggested that environmental learning might serve purposes of raising awareness, prompting conceptual or behavioural change, promoting moral understanding and developing metacognitive skills. Others emphasise intellectual developments alongside more activist outcomes, suggesting the purpose is to enhance students' knowledge and critical thinking about the environment and society so as to enable them to participate and take action as both local and global citizens, voters, and consumers. Such ideas have, for example, informed Jensen and Schnack's (1997) notion of action competence.

Our approach takes these different views into account, but adds the crucial perspective of the learner. Much of the literature discusses why environmental learning is important, why we should encourage and support people in such learning. Our phrasing is deliberate: why *they* are learning. This prompts us to explore why students engage in learning, what purpose or value they perceive in their learning. In our research we did account for teachers' objectives in setting up particular learning experiences, but our focus was on the purposes of and motivations for learning perceived and expressed by learners themselves. These emerge in many of the discussions throughout later chapters. Both within settings and between them we found considerable diversity in students' views as to the purpose of their learning, and to the reasons why they thought learning was worth pursuing. These included many of those discussed above: learning that serves vocational futures, learning that contributes to wider social understanding of important issues, learning that supports intra- and inter-personal understanding of environmental values, learning that leads to behaviour change, and learning for the sake of knowledge and understanding in and of themselves.

Having explained each of the five segments of Fig. 1, we now offer three examples, drawn from our studies, which illustrate how these may come together in different combinations (see Box 2).

Box 2 Examples illustrating variation in the who, where, what, how and why of environmental learning

Example 1
Swedish undergraduates studying for a degree in engineering took an Ecology course, a component of which focused on biogeochemical cycles (a focus on nature, or 2 in our model). The primary means of learning was through a series of lectures, and the purpose was seen (among other things) as developing 'basic scientific knowledge'.

Example 2
Swedish undergraduates in the discipline of biology took a course on environmental auditing, 'Environmental control for biologists', within a biology department. The subject matter focused on the relation between society and nature (5 in our model), and was pursued through open-ended group and project work. Students described purposes relating to learning knowledge and skills relating to environmental auditing, but also developing critical perspectives on related issues.

Box 2 (continued)

Example 3
English secondary school students attended (compulsory) geography lessons focused on fair trade. These explored relationships between individual behaviour and society (6 on our model), and students engaged in a range of tasks (drawing diagrams, role-playing different characters involved in trade, and empathetic writing) which were set up by the teacher. Students from the same class had differing views as to the purpose of this learning: some felt there was an implicit message that they should buy fair trade products (perceiving behaviour change as intended aim), while others felt it was more about promoting cross-cultural understanding.

Summary

In this chapter we have provided a brief outline of aspects of the wider literature which are most pertinent to the argument we develop through this book. In particular we have emphasised the breadth of environmental learning in terms of its possible foci, outcomes, contexts, processes, qualities and underlying assumptions. We then offered a framework which describes the way we have come to work with the notion of 'environmental learning', highlighting five aspects which we suggest are important to consider: who, what, where, how, and why. Under each of these headings we have shown how the focus of this book relates to the phenomenon of environmental learning more broadly. In the concluding sections of each of the later substantive chapters titled "Dealing with Emotions and Values", "Questioning Relevance" and "Negotiating Viewpoints Among Students and Teachers" we use this framework to reflect on what we have learned by highlighting what our studies reveal about the who, what, where, how and why of environmental learning. Our next concern, though, is to consider how this complex phenomenon of environmental learning has been approached by researchers.

Researching Environmental Learning

Having explored the nature and characteristics of environmental learning, in this chapter we move on to consider the ways in which researchers have sought to explore different aspects of this topic over recent years. We identify three contrasting research approaches within the literature (researching learners, measuring outcomes, exploring processes) and discuss recent work on students' environmental learning experiences as one example of the third category. The chapter ends with a brief overview of what we know about learners and learning in environmental education within formal settings.

Environmental Learning – An Emerging Research Topic

The first point to make about researching environmental learning is that it has only fairly recently emerged as a topic of concern amongst environmental education researchers. As noted in our opening chapter, environmental learning has been a topic that has been noticeable by its absence within the research journals in environmental education for much of the last decade.

Writing in the late 1990s, Payne (1998, p. 20) drew attention to 'a lack of consideration in environmental education theory and research practices about the children who are the subjects of environmental education'. We have argued elsewhere (Rickinson 2006) that this 'lack of consideration' has been reflective of three underlying weaknesses in the field of environmental education research. First, there has been a tendency to neglect learning both in a literal sense (i.e. there have been few empirical or theoretical studies) and in a more subtle sense (i.e. where learning has been studied, it has tended to be in relation to educational interventions rather than as a process in its own right). Secondly, concerns about environmental learning *processes* have been nearly completely overshadowed by interest in environmental learning outcomes. Much has been written about the impact of educational programmes in terms of pre-determined learning outcomes, but there have been few attempts to research or conceptualize the processes through which environmental learning might be taking place. Thirdly, as emphasized by Dillon (2003),

M. Rickinson et al., *Environmental Learning: Insights from Research into the Student Experience*, DOI 10.1007/978-90-481-2956-0_3,

there has been an apparent reluctance amongst environmental education researchers to engage with learning theory. This is in stark contrast to a field such as science education where 'learning theories are much more part of the discourse of significant numbers of researchers' (p. 217).

This situation, however, is not static and there are a number of signs of positive developments. As we noted earlier, there is growing recognition of the centrality of learning within debates about sustainability (e.g. Sterling 2001; Scott and Gough 2003). There are more publications emerging with a focus on learning in environmental and sustainability education (e.g. Heimlich 2005; Hart 2007; Wals 2007; Reid et al. 2008; Stevenson and Dillon forthcoming) and a recent publication reflecting on research in education and the environment from 1996 to 2006 featured 'environmental learning as process and outcome' as one of six themes (Reid and Scott 2006a).

Other encouraging signs include the increasing number of studies that take seriously the ways in which learners are making sense of their experiences of environmental and sustainability education (e.g. Lai 1999; Ballantyne and Packer 2002; Nagel 2004). There is also evidence of efforts to explore how more general theories of learning might inform research in environmental education. Dillon (2003), Falk (2005) and Myers (2006) provide overviews of various perspectives on learning and advocate further work on models of learning in the context of environmental education. Meanwhile, Reid and Nikel (2008) have examined how different perspectives on learning might inform alternative framings of participation in environmental education. There are also examples of empirical studies that have discussed their findings in relation to particular theories or models of learning (e.g. Brody and Tomkiewicz 2002; Österlind 2005).

Ways of Researching Environmental Learning

Notwithstanding the fairly recent emergence of research focused specifically on environmental learning, it is possible to identify a number of different ways in which researchers have sought to investigate this theme. Distinctions can be drawn between studies that have sought to generate:

- Information about learners (researching learners)
- Evidence on learning outcomes (measuring outcomes)
- Insights into learning processes (exploring processes) (Table 1).

As with all such simple categorisations, its distinctions are not water-tight but rather reflective of differences in emphasis. In other words, there is considerable potential overlap between the categories, which need to be seen as a helpful summary (rather than a definitive classification) of possible approaches to researching environmental learning.

Table 1 Different approaches to researching environmental learning

Type of research	Typical foci	Typical methods	Examples
Researching learners	Students' environmental knowledge, perceptions, attitudes or behaviours	Survey (with some interview or mixed-method)	Kuhlemeier et al. (1999)
Measuring outcomes	Impact of programmes on students' environmental knowledge, perceptions, attitudes or behaviours	Pre-test/post-test survey (and some interview)	Bogner (1998)
Exploring processes	Students' experiences of, and responses to, different kinds of environmental learning	Interview, observation	Lai (1999)

Researching Learners

This first category comprises studies that have sought to discover more about some aspect of young people's environmental knowledge, attitudes and/or behaviours. Studies in this category have, therefore, not specifically been about understanding environmental learning itself, but rather investigating the environmental characteristics of young people. The sorts of issues that have been studied have included:

- Young people's factual knowledge about environmental phenomena
- Students' understanding and misunderstanding of environmental phenomena
- The sources of young people's environmental information
- The kinds of environmental issues that students are concerned about
- The extent and nature of students' environmental concern and/or indifference i.e. their environmental attitudes
- The extent and nature of students' involvement in environmentally responsible practices, i.e. their environmental behaviours
- The kinds of factors that are influential upon young people's environmental knowledge, concerns, attitudes and behaviours.

Across all of these various topics, there has been a stronger focus on quantitative survey-based methods as opposed to qualitative interview-based approaches. Most studies have involved questionnaire surveys (both fixed- and open-response) administered to particular samples of young people in order to assess the *extent*, rather than the *nature*, of students' environmental knowledge, attitudes and behaviours (see, for example, Box 1). That said, there have been some examples of qualitative inquiries that have sought to understand children's perspectives through techniques such as open-ended interviewing (e.g. Cullingford 1994; Palmer 1995).

Studies of learners' environmental knowledge have often been stimulated by the view that greater awareness of learners' pre-existing ideas (particularly those in conflict with 'present scientific understanding') should be helpful in designing more effective teaching strategies (Boyes and Stanisstreet 1994, p. 313). Attitude and behaviour studies, meanwhile, have typically been motivated by arguments such as: (i) an understanding of environmental attitudes is crucial because 'they are a precursor

Box 1 Example of a study focused on 'Researching Learners'

Kuhlemeier et al. (1999) undertook a study of over 9,000 secondary school students (aged 15 years) from 206 schools across The Netherlands. Using a questionnaire survey, the research team investigated the young people's environmental knowledge, attitudes and behaviours. The survey assessed the extent of agreement or disagreement with statements such as: 'Owing to pollution of the environment, the world threatens to become unfit to live in for future generations'. It also asked questions about specific behaviours, such as 'How often do you use more tap water than is strictly necessary, e.g. take a long shower or let the water run continually when brushing your teeth or doing the dishes?'.

The main findings were summarised as follows:

- Fifty-seven percent of the students had a (very) positive attitude toward the environment, and 35% were prepared to take extra pains or to make (financial) sacrifices for the environment.
- However, the students' knowledge about environmental problems was fragmentary and often incorrect. Similarly, the environmentally responsible behaviour of many of the students was inadequate.
- The relation between environmental knowledge and environmental attitudes and behaviour proved to be very weak.
- Consistent with theories on attitudes, environmentally responsible behaviour was more strongly connected with willingness to make sacrifices than with attitude toward the environment.

Source: Kuhlemeier H, Van den Bergh H, Lagerweij N (1999) Environmental knowledge, attitudes, and behavior in Dutch secondary education. J Environ Educ 30(2):4–14.

to pro-environmental behaviour, which is the ultimate goal of environmental education' (Chan 1996, p. 298); and (ii) young people hold environmental attitudes and 'to achieve success in environmental education teachers need to identify and draw on children's opinions about environments' (Kwan and Miles 1998, p. 12).

Measuring Outcomes

This second category moves beyond studies of the characteristics of learners into research on learners in relation to environmental education programmes. The focus here is on the outcomes of educational interventions in terms of the extent to which they bring about changes in students' environmental knowledge, attitudes and/or behaviours (see Box 2 for an example).

Such studies are almost universally ones that seek to evaluate the effects of particular educational treatments (independent variables) on students' environmental

Box 2 Example of a study focused on 'Measuring Outcomes'

Bogner (1998) investigated the effects of a 5-day residential outdoor ecology programme in a German national park.

Aimed at secondary school students, this programme involved students working in small groups undertaking a mixture of 'cognitively-based activities' such as locating and identifying animal tracks, and emotionally-orientated ones such as 'Touch a Tree' and simulation games. The aim was to provide an 'original encounter with biological and ecological themes [through] structured and participatory learning activities on trails within a woodland area'. There was also an onus on avoiding 'the kinds of partitions and divisions that are found in normal biology syllabi' (p. 20).

Through comparison with two control groups that either participated in an outdoor programme without an ecological focus or did not attend any programme, this study reported the following effects 1 month after the programme amongst the sample of 351 secondary school students (12–13 years):

- Significant gains in students' environmental knowledge
- Significant changes in students' attitudes towards human utilisation of nature
- Significant shifts in students' willingness to plan and take action for the environment.

The key conclusion of this work was that participation in this well-established outdoor ecology programme did affect positive changes in students' environmental knowledge, attitudes and behaviour.

Source: Bogner FX (1998) The influence of short-term outdoor ecology education on long-term variables of environmental perspective. J Environ Educ 29(4):17–29.

knowledge, attitudes or behaviour (dependent variables) through some kind of quasi-experimental pre-test/post-test design. Much of the evidence that has been generated on outcomes is quantitative in nature and based on outcome criteria that are tightly specified prior to the investigation. In other words, the research questions being asked in this category are of a closed nature such as 'To what extent does treatment x change students' attitudes towards y?' There are very few examples of outcomes studies that start out with an open question such as 'What impact does initiative x have upon y students?' (Emmons 1997 is an example.)

Studies of learning outcomes have been motivated by two main arguments: there are many possible forms of environmental education, and 'evaluation is important to determine which forms are most effective' (Dettmann-Easler and Pease 1999, p. 33); and changes in individuals' environmental knowledge, attitudes and behaviours are crucial goals for environmental education, and so are appropriate criteria for programme evaluation (e.g. Bogner 1998 p. 18).

Exploring Processes

This final category, which has become more prominent over recent years, is focused on exploring the processes of environmental learning as opposed to its outcomes. There is a strong emphasis in this work on the voice of the student and seeking to understand how learners themselves make sense of environmental education. Studies work from a conceptualisation of learners as active agents, rather than passive subjects, in the learning situation. In keeping with this, most studies have been qualitative in nature, involving semi-structured interviews and participant observation of learners (see Box 3). There has also been some survey-based work on, for example, students' preferences for environmental education (Roper Starch Worldwide 1994).

Research topics within this category have included:

- Students' memories and views of their school-based environmental education (e.g. Roper Starch Worldwide 1994; Battersby 1999; Connell et al. 1999)
- Young people's experiences of particular environmental learning situations (see below for examples and more detailed discussion)
- Adult learners' responses to and experience of non-formal environmental education through, for example, national park visits (Brody and Tomkiewicz 2002).

The sorts of arguments that have underpinned these types of studies have been that: (i) the voice of the learner is a severely neglected one in environmental education research and curriculum development (Payne 1998; Rickinson 1999b; Nagel 2004); (ii) understanding learning necessitates a 'focus on the multiple ways in which [learners] make sense of the information they encounter, rather than whether [they] "get the message" the provider intended to convey' (Ballantyne and Packer 2005, p. 283); and (iii) learning is an active process influenced by emotional factors, and needs to be researched as such through naturalistic consultation with students about their experience and learning (Lai 1999).

One strand of work within the 'exploring processes' category has been studies of young people's responses to and experience of various forms of environmental learning. This is the area of the field that best encompasses the research that we are reporting in this book. As well as our studies (described in the Introduction and Appendix I: Empirical Context), it also includes work by others on young people's responses to field trips (Lai 1999), nature-based excursions (Ballantyne and Packer 2002), secondary school science classes (Österlind 2005), primary school environmental education (Nagel 2004; Nagel and Lidstone 2008), and school-based social/environmental change programmes (Mckenzie 2006). What's distinctive about these kinds of studies is their concern with the way environmental curricula are experienced by learners in the everyday contexts of real-life learning settings. As such they are interested in students' affective responses to the curriculum, their varied interpretations of learning tasks and the kinds of learning strategies and processes that they use. In short, work in this area is about taking seriously 'the subjective experience of students as they are engaged in learning' (Erickson and Shultz, 1992, p. 466).

Box 3 Example of a study focused on 'Exploring Processes'

As part of a bigger project on geography fieldwork in Hong Kong, Lai (1999) investigated the fieldwork experiences of three groups of secondary school students (15–16 year olds) from a boys' school in Hong Kong. The study was undertaken in response to the author's view that 'studies of the actual teaching learning process in fieldwork, especially how teachers and students make sense of their experiences, are generally neglected' (p. 242).

The study used a multi-method approach, collecting qualitative data through interviews, participant observation and documentary analysis. Three geography teachers and three groups of students were interviewed separately before and after the field trip. The researcher also took part in the field trip as a participant observer and took field notes and photographs. The worksheet used during the trip and the students' post-trip projects were also studied.

The results revealed that the students' intent before the trip and their learning experiences after the trip were much richer than their teachers had intended. Despite some students' negative experiences of previous field trips, there was a universally strong desire to escape from their perceived boredom and constraints of the classroom.

The study revealed the relative freedom of the field as a learning environment in which the students were more proactive and teacher-student rapport improved. Besides deepening their understanding of what they had previously learned in class, some students were able to see things from a new perspective.

The findings also indicated that upon return to school, there was a quick return to the status quo. The inspiration which resulted from freedom of learning from the field trip was often not transferred to the classroom setting.

Source: Lai KC (1999) Freedom to learn: a study of the experiences of secondary school teachers and students in a geography field trip. Int Res Geogr Environ Educ 8(3):239–255.

Current Knowledge About Environmental Learning in Formal Settings

Taken together, what does the evidence stemming from these three categories of research tell us about environmental learning within formal education? Drawing on an earlier synthesis of relevant research (Rickinson 2001), we can answer this question in terms of information about learners and information about learning.

Learners in Environmental Education

The current evidence provides information about two aspects of learners in relation to environmental education: their environmental characteristics (such as attitudes, knowledge, concerns, perceptions, etc.) and their educational perspectives (such as experiences, preferences, etc.). A recurring theme throughout all of this evidence is that students are not neutral or 'blank slates' in relation to either the environmental or the educational aspects of environmental education. Studies of students' environmental attitudes, concerns, perceptions, and understandings, for example, generate pictures of young people having 'robust models' of global environmental phenomena (Boyes and Stanisstreet 1997), deeply held feelings (including fears) about natural areas (Simmons 1994; Wals 1994), and particular views about the future and perceptions of nature (Hutchinson 1997; Bonnett and Williams 1998; Payne 1998). Meanwhile, research on learning experiences and preferences report varying perspectives on learning through fieldwork (Lai 1999), classroom learning tasks (Schindler 1993; Rickinson 1999b), university environmental courses (Lundholm 2005), environmental information sources (Connell et al. 1998), the nature of school-based environmental education (Connell et al. 1999; Nagel 2004), and outdoor activities (Bixler and Floyd 1999).

Several studies also highlight the intricate and often contradictory nature of students' perspectives. It is clear that the views of students are complex and varied, rather than straightforward and simple. Bonnett and Williams (1998) and Payne (1998), for example, both describe groups of children expressing contradictory or ambivalent perceptions of nature, as something both separate from, but then also including of, people and human-made objects. Studies of attitudes, meanwhile, note conflicts between a pro-environmental stance on many issues, but a less environmentally conscious attitude about issues linked with their own lives and material aspirations (Chan 1996; Connell et al. 1998; Ivy et al. 1998). Similarly, in studies of classroom learning (Rickinson 1999b; Hopwood 2007a) and fieldwork activities (Lai 1999), a marked variety is found in individual students' perspectives on, and responses to, particular aspects of their learning situations.

The origins and development of students' environmental and educational perspectives are not well understood. What is known is that environmental characteristics can (i) vary with demographic factors such as gender, age and socio-economic grouping; (ii) change at least in the short term as a result of educational interventions; and (iii) be affected by influences such as the media and schooling. The way in which these influences play out over time, though, is currently unclear, due to the fact that our current understanding of learners is a largely static one. In other words, we do not know a great deal about how students' perspectives develop and/or change over time.

Learning in Environmental Education

The current evidence base provides information about two aspects of environmental learning: learning outcomes (usually of groups of students), and learning experiences (usually of individual students).

The evidence on learning outcomes shows that environmental education programmes can affect change in students' environmental attitudes, knowledge and (in a few cases) behaviours. The evidence relates to a variety of both school-based and outdoor educational initiatives. Effects, however, tend to be measured in the short term, and in most cases their durability over time is unclear. Furthermore, positive effects can be partial in that change in attitudes – though not behaviour – might occur, or can be absent altogether. It is not well understood how or why particular outcomes do or do not occur, although there is some evidence to suggest that certain aspects of programmes are helpful in yielding positive impacts. These aspects include: programme duration in terms of week-long, as opposed to shorter, outdoor field courses (Emmons 1997; Bogner 1998); preparatory and follow-up work in connection with excursions to local zoos/museums (Gutierrez de White and Jacobson 1994; Farmer and Wott 1995); parental and community involvement with programmes (Uzzell et al. 1994; Milton et al. 1995; Ballantyne et al. 1998); and authenticity of content in terms of dealing with actual, local environmental issues (Uzzell et al. 1994). There is also some evidence to suggest that learning outcomes can be facilitated by certain processes such as role modelling and direct experience aiding attitude change on outdoor courses (Emmons 1997) and collaborative group discussion helping conceptual development in classroom lessons (Mason and Santi 1998).

In addition to the influence of various programme characteristics, evidence suggests that learning outcomes can also be affected by the nature of the participating students. A small number of studies have found variations in outcomes in relation to: students' age and level of interest in the environment (Leeming et al. 1997), students' ability levels (Gayford 1995), and the extent to which students enjoy a programme (Ballantyne et al. 1998). This concurs with emerging evidence about students' learning experiences, which indicates that individual students can make sense of similar learning tasks in quite different ways. What is clear from this is that students are active experiencers, rather than passive recipients, of environmental curricula, and respond to learning situations in individual ways.

In summary, it would seem that the evidence from studies of learning outcomes and from studies of learning experiences are both highlighting a common point – the importance of the learner in the process of learning. It is with this in mind that we can move towards a more detailed discussion of what the learners in our research studies told us about their encounters with environmental learning activities within secondary school lessons and university courses. To this end, the next chapter introduces three lenses that help to highlight the three aspects of environmental learning that were significant for the students within our studies.

Summary

Until recently, there have been few empirical investigations specifically focused on learning in environmental education. The tendency has been to focus on environmental learning outcomes as opposed to environmental learning processes and to make little use of wider learning theory. This situation is changing, however, as research in environmental learning grows and diversifies. Three broad research approaches can be identified within the literature: researching learners (focused on students' environmental knowledge, attitudes or behaviours); measuring outcomes (focused on the impact of programmes on students' environmental knowledge, attitudes or behaviours); and exploring processes (focused on students' experiences, processes and responses to different kinds of environmental learning). The research discussed in this book is illustrative of the third of these categories concerned with the process aspects of environmental learning. Across all three research approaches, however, we see evidence highlighting the importance of the learner in the process of learning. To understand in more depth what this means in the context of everyday school and university classrooms, the next chapter introduces three lenses that have helped us to better understand the aspects of environmental learning that were significant for the students within our studies.

Lenses for Understanding Environmental Learning

This chapter introduces the idea of lenses – interpretive tools or heuristic devices which lead us to look at or for different aspects of environmental learning and thus reach qualitatively different understandings. It also outlines the three lenses that we have developed and which we apply in subsequent chapters to make sense of empirical data across a range of formal learning settings. We begin by explaining what we mean when we use the word *lens*, and this is followed by a hypothetical environmental learning scenario used to illustrate the different lines of enquiry framed by each lens. We then explain their origins, provide a rationale for their importance, and conclude with a schematic representation and summary.

Introducing the Concept of Lenses

A key concept in this book is that of *lenses* – conceptual devices which can be used to understand environmental learning in qualitatively different ways. We have developed three distinct lenses as a result of an extended period of collaborative data sharing and analysis. These are introduced later in this chapter and are discussed in depth in the chapters titled "Dealing with Emotions and Values", "Questioning Relevance" and "Negotiationg Viewpoints Among Students and Teachers". Before we explore these different ways to make sense of environmental learning, we will explain in more detail what we mean by a lens.

Gee (2001) describes a 'set of interpretative tools that lead us (like all interpretive tools) to look more closely at certain issues and less closely at others' (p. 100). The lenses we develop and use in this book are likewise heuristic devices which suggest looking at or for certain aspects of environmental learning, and divert our gaze away from others. We also follow Wenger (1998) in presenting these lenses not as recipes which guarantee particular outcomes, but rather as guides about what to pay attention to. We see such lenses as different from theories in that they do not in themselves make claims about the world, but rather provide a frame for researchers in making choices about what to look at or what to look for in their analyses and interpretations.

Each lens points to a series of questions, or lines of inquiry that concentrate thinking on certain issues. Such guiding is limited in scope – their bounded nature is

M. Rickinson et al., *Environmental Learning: Insights from Research into the Student Experience*, DOI 10.1007/978-90-481-2956-0_4,

what makes them practical, accessible, and functional. None can account for all aspects of environmental learning, and the three we present in this book draw attention away from many aspects by virtue of their leading us to look at and for particular things. The lenses do not offer a set of predictive or generalisable claims or principles, but offer different vantage points from which to make sense of environmental learning. They delimit areas of focus, pose particular questions, and draw on other relevant ideas, concepts and theory.

But why *lens*? In many ways optical lenses, such as those found in a microscope, are like those in this book. When using a microscope, one can choose to look at whatever is on the slide through a number of different lenses. Each offers a very different view of the same object, but none provides an exhaustive or complete view, and none is 'truer' than the other. They may be differently useful, depending on what one is looking at or for. Some features of the object will be apparent through several lenses, while others appear sharper and are best explored through only one. Each reveals certain features that are otherwise hard to see and 'hides' others. The microscope itself does not determine the nature of knowledge which is produced through its use; it remains for the user to interpret what is seen, often drawing on existing knowledge and understanding.

In this book the object under examination is environmental learning in formal contexts. The three lenses we describe and apply enable us to paint three different pictures of environmental learning. Their differences reflect variation in purpose, lines of questioning and noticing. The lenses provide a framework for *looking at or for* and help us interpret what we see by pointing to relevant ideas and concepts from other work.

Thus, in this book we use the metaphor of a lens in reference to an heuristic device which…

1. Foregrounds particular issues as warranting our attention.
2. Sets out a line of inquiry or set of questions to pursue.
3. Diverts attention towards particular things as well as away from others, thereby offering a manageable conceptual realm to deal with.
4. Points to relevant theoretical and empirical work.
5. Stresses (through being one of many possible alternatives) that there is no one correct or exhaustive view or understanding of environmental learning.
6. Through all of the above, encourages a diversity of standpoints and provides a means to look across multiple studies and areas of literature.

We illustrate these characteristics and introduce the particular qualities of the three lenses proposed in the next section.

Introducing the Three Lenses

The purpose of this section is to introduce the three lenses that arose from our collaborative data sharing and analysis, discussed in greater detail in subsequent chapters. They focus on emotions and values (lens 1), issues to do with relevance (lens 2), and differing viewpoints among students and teachers (lens 3).

In order to illustrate the application of different lenses to a common context – emphasising the idea of seeking qualitatively different understandings of a shared phenomenon – we present an hypothetical scenario of environmental learning in a formal setting. This is then used as a point of reference in the outlines of each lens which follow.

An Exercise in Imagination

The learning experiences we imagine are set in a formal context (consistent with the focus of this book) but beyond this remain deliberately footloose. The description provided in Box 1 makes no specific reference to the age of the learners, geographical location, or educational setting. Such an account might be generated with reference to school age or university students, in a number of countries, and in the context of several subjects or courses. It is however explicitly framed around formal learning experiences.

Depending on our purposes, the questions we ask, the ideas, concepts, and theoretical perspectives we draw upon, we will notice, value, and attend to different aspects of these learning experiences. The following paragraphs continue the hypothetical exercise to consider three different lenses through which one might understand what is happening in the class on energy policy.

Lens 1: Focus on Emotions and Values

In seeking to understand what was happening and why in this class, we might focus our attention on values and emotions. Environmental learning (as we discussed in the chapter titled "What Is Environmental Learning?") frequently involves confronting controversial issues. The very nature of what environmental learning is about can evoke emotional responses, and can foreground the different values held by learners and their teachers. We can imagine a two-way interaction between learners and what is being learned: the focus of environmental learning can provoke emotional reactions in students, and at the same time a student's own values and emotions can mediate their response to particular learning experiences. We could explore different affective responses to the content and issues being discussed: How did students *feel* about nuclear power, pollution? Whose values were brought to bear in the lesson, and by whom? Did they experience emotional conflict when engaging in the role play exercise? How did they perceive the relationship between evidence, argument and opinion in their final written assessment piece? Our first lens sets these kinds of questions in a broader context, and points to related empirical and theoretical work which informs looking at or for emotions and values as they play out in the process and experience of environmental learning.

Box 1 Hypothetical environmental learning scenario

A group of learners arrives at a pre-specified time and place to attend a class. They have diverse backgrounds, interests, identities and expectations about the learning they expect will take place. The focus of today's learning is introduced by the teacher: the future of national energy policy.

Initially, the teacher chairs an open discussion in which students are asked to suggest issues and ideas which might be relevant to this theme. This builds on previous learning experiences in which different energy sources and issues of energy security, sustainability and environmental impact had been studied. This class concludes the topic by debating the direction in which learners feel their country's energy policy should go.

Students are then asked to get into small teams, and each is assigned a particular group of people to represent in a role-play scenario: an environmental lobbying organisation, a charity which supports economically deprived people in inner cities, the government ministry responsible for energy, a large multinational petroleum company, and residents of a rural community located near to a major piece of energy infrastructure. The remit is to prepare for a panel-style debate in which different energy interests and priorities will be considered. Several students initially protest at the group they have been assigned to, but the teacher insists that no changes will be made, and that the students should focus on developing the strongest possible argument that they think would reflect the interests and views of their nominated group.

The classroom is soon permeated by lively conversation and an atmosphere of busyness among the students. The teacher walks around the room and listens to the group conversations. One team spends some time discussing what each person will do, assigning each member responsibility for one particular argument. Another brainstorms possible arguments for and against their point of view, two struggle to agree on what the perspective of their assigned group would be, and others involve a mixture of activity focused on the task with less relevant chatter.

The teacher calls the class to attention, and each team is given 5 minutes to make its case in front of the class. These form the basis for a lively whole class discussion in which the teacher acts as a neutral chair and tries to ensure that not only the voices of more vocal and confident students are heard.

The final segment of the class is devoted to individual written work in which students are asked to explain how they think national energy policy should be developed, taking into account the support and contestation their particular view would be likely to meet from different groups. This task is to be finished in the students' own time and handed to the teacher for assessment purposes. Soon after the students begin this work, several initiate dialogue with the teacher asking, for example, whether it matters which perspective they choose.

Lens 2: Focus on Issues To Do with Relevance

We would also see value in foregrounding issues to do with relevance. Environmental educators, like many others, are frequently called upon to ensure that environmental learning is perceived by learners as being relevant to their own lives. The students judge the relevance of the experiences described in Box 1 according to their own past experiences or present context. It could conceivably make an important difference if this debate were to be held among students with different personal investiture in energy issues – family ties to employment in fossil fuel production, living near sites of proposed new nuclear power plants and so on. Some students might feel disempowered in relation to affecting national energy policy and deem the learning interesting but not of real-life relevance. Others may see these issues as of great practical and political importance in their own lives now and in the future.

Students in this class might also be interpreting these learning experiences in terms of their relevance to the curricular context in which they occur. If this were part of a secondary school science lesson for example, evidence from existing research suggests that some students might find the controversial and opinionated nature of the content difficult to square with their image of science as objective and factual in nature (Driver et al. 1996). Different perceptions of relevance might be brought to bear if this learning was part of a course on economics, politics, or geography. It might be anticipated that such evaluations as to the relevance of learning might influence the nature of student engagement in learning experiences, and thus the process and outcomes of learning. Our second lens develops these ideas further, pointing to the wide range of ways in which the notion of relevance has surfaced in our research and in the literature.

Lens 3: Focus on Differing Viewpoints Among Students and Teachers

This class can also be seen as a collective social setting, comprising different roles (teacher, learner) but also variation within these (differences among learners). One could explore how viewpoints among this group (including the teacher) vary, and what the consequences of conflicts and differences of opinion might be. These may not relate only to the subject matter (in this case views on appropriate forms of energy policy), but may also relate to views about the nature of the learning activities (students complaining to the teacher when they were assigned the group to represent in the role play), or about what is required to do a task well (concerns among students that they may earn fewer marks in their written work if they express a view which is not congruent with that of the teacher). These issues are explored in more detail through our third lens, which examines environmental learning situations in terms of the interactions between the viewpoints of students and the viewpoints of teachers. Informed by wider research on the curricular dimensions of student–teacher interactions/relations, this lens draws attention to the ways in which differences between students' and teachers' perspectives can present challenges for student engagement and learning.

How the Lenses Came About

In this section we provide a brief account of where these lenses came from. A fuller methodological account is provided in Appendix II: Development of the Lenses, and has also been presented by Lundholm et al. (2008).

Each lens took embryonic form in the studies we conducted independently (Rickinson 1999a, b; Lundholm 2003, 2007; Hopwood 2006, 2008). Over a period of 4 years they were developed through the sharing of raw data and exploration of the extent to which the issues arising from one study were evident across the other two datasets (and the different contexts and educational settings in which they were situated). The chapter titled "Introduction" provides details of the three empirical studies.

The sorts of questions posed, aspects of environmental education focused on, and concepts or theoretical work drawn upon reflect in many ways the characteristics of our initially separate research agendas. Over time we have followed our own courses in making sense of each other's data and findings reported in the wider literature. Through this process the three lenses have been distilled and refined. Remnants of our own initial lines of enquiry can be traced in the shape of the lenses as presented in this book. By looking across several studies, new questions, ideas and concepts have been added which do not remove all trace of earlier thinking, but rather constitute an additional layer and contribute to an increasingly complex and rich picture.

This process started with an initial recognition of similarities between the findings of our separate studies (see, for example, Lundholm and Rickinson 2005, 2006). However, the lenses do more than merely identify parallel outcomes: they offer qualitatively different ways of seeking to understand environmental learning and a means to pursue such lines of enquiry across multiple datasets.

Why These Lenses Are Important

We will now outline arguments as to why it is important to look at or for values and emotions, issues of relevance, and differing viewpoints between teachers and learners. These discussions also reference some of the key ideas and areas of theoretical and empirical literature that were pointed in the introductions to each lens provided above.

The Importance of Values and Emotions

The role of emotions in learning has received increasing attention in educational research in recent years (Pekrun 2005; Sansone and Thoman 2005). Emotional aspects of learning include motivation and students' engagement in activities, whether in schools, universities, outdoor sites or other non-formal learning settings. Affective factors, such as interest, can be seen as a 'driving force' in learning, leading students towards more meaningful engagement rather than rote memorisation (Pintrich et al. 1993; Dole and Sinatra 1998). Evidence suggests conversely that students may disengage from learning activities and tasks if they experience dislike or discomfort with what is being learned (Watts and Alsop 1997). This is crucial when understanding students' emotional reactions to contested and highly charged issues such as climate change. Emotions are also cited as 'guiding' lights affecting learning through directing students' interest and attention (Claxton 1989; Lundholm 2005; Österlind 2006).

Furthermore, there is growing interest in understanding the way emotions and values are important not only in influencing motivation and engagement at a general level, but also in shaping the process of conceptual development and change (Pintrich et al. 1993; Sinatra and Pintrich 2003; Sinatra 2005; Watts and Alsop 1997). Thagard and Zhu (2003), for example, argue that:

> People do not only hold and use their concepts and hypotheses, they also feel emotionally attached to them and respond with negative emotions to concepts and hypotheses that clash with them ... Having the intention to understand and evaluate alternative views can make the emotional component of conceptual change more easily realized. (p. 98)

Mason's (2003) review of research on students' epistemological beliefs suggests not only that such beliefs may strongly influence learning, but also that values and emotions play a similarly important role. For example Schommer (1990, 1993) has shown that students' beliefs about the nature of knowledge (simple or complex, certain or uncertain) appear to be related to their responses to different kinds of subject matter, as when one encounters evidence that conflicts with one's beliefs. Mason (2003) makes explicit the link between personal conceptions and dealing with varied opinions or controversies, suggesting more effective reasoning and critical thinking may be facilitated by an openness to suspend the former and examine the latter. It should be remembered, however, that students' epistemological beliefs and responses to subject matter may at least in part reflect context (see Leach et al. 2000) – hence our attention to the setting and detail of learning experiences at hand in the chapter titled "Dealing with Emotions and Values".

Studies working with notions of scientific literacy have found that learners may distinguish between the subjectivity of opinion and the objectivity of scientific knowledge (Zeidler et al. 2002), and may use different kinds of knowledge alongside values and emotions in decision-making on socio scientific issues (as in Sadler and Donnelly's (2006) study which explores genetic engineering). The fact that environmental issues are value laden and emotionally charged, as discussed in the chapter titled "What Is Environmental Learning?" is highlighted and addressed by looking through this lens. There is strong evidence across a significant body of literature that emotions and values are an important aspect of learning. This lens brings into focus students' emotions, engagement and disengagement, values and views of subject matter specifically in relation to environmental learning in formal contexts.

The Importance of Relevance

Learners' perceptions of the relevance of their learning have been of interest to researchers for some time, across a range of subject areas in both school and higher education contexts. Reviews of research on pupils' experiences of learning under the National Curriculum in England (Lord and Harland 2000; Lord 2001, 2002, 2003; Lord and Jones 2006) have found relevance to be a commonly investigated theme. However, as Lord and Jones (2006) explain, relevance has a diversity of meanings and has been explored in different ways. They distinguish research on academic relevance (passing exams, getting grades), vocational relevance, usefulness to current needs and personal life, relevance to future and adult life, and connections with real life. In our use of relevance as a lens through which to understand environmental learning, we differentiate (i) relevance to the learner, whether in the present or future, and (ii) relevance to curricular context. The latter is perhaps a less common notion of relevance, but is one that we have found to be important in the experience of environmental learning as evidenced in our studies.

Students are more likely to engage in, value, and learn from learning experiences if they are perceived to be relevant in some way or other. Postman and Weingartner (1969) suggest that researchers and teachers should examine learning experiences by asking "where is the learner in all of this? Where is his [sic] world?" (p. 43). They continue:

> There is no way to help a learner to be disciplined, active, and thoroughly engaged unless *he* [sic; emphasis in original] perceives a problem to be a problem or whatever is to-be-learned as worth learning… unless an inquiry is perceived as relevant by the learner, no significant learning will take place. No one will learn anything he [sic] doesn't want to know. (p. 52)

A commonly discussed and reasonably intuitive idea is that learners divest effort into learning and enjoy it more if they see it as relevant to their own contexts and everyday lives. Beane (1997) and Nagel (1996) contend that learners find their work more interesting and the content more relevant if the (at least initial) focus is on real life issues; if this condition is met, such arguments suggest, then students are more likely to be motivated to achieve deep understanding rather than surface engagement with learning material. Such a claim is supported by evidence from studies in several subject contexts (Lord and Jones 2006). According to Pintrich et al. (1993) motivational beliefs are important mediators of classroom learning.

Motivation, they suggest, has several dimensions including personal interest, utility value, and importance – all of which connote ideas of relevance. They distinguish (as do we) relevance to the present and relevance to future (career) selves. They establish the importance of learners' perceptions of 'the salience or significance of the content or task to the individual' (p. 183).

Williams and Kalvin (2007) argue that if environmental education is to be effective, it must be relevant to its constituents (i.e. learners). Again referring to environmental learning Connell et al. (1999) note among Australian learners a 'desire for practicality and relevance in their studies' (p. 103; see also DiEnno and Hilton 2005). Barratt and Barratt Hacking (2008) respond to findings from their research with English schoolchildren with a call to localise environmental curriculum content in order to make it 'relevant to children's everyday life' (p. 292). Lynch (2003) and Brooks (2003) argue that the school subject of geography should be made relevant to young people by focusing on current issues and events, tapping into what learners themselves are interested in. Twenty-five years ago Knamiller (1983) argued that environmental education was often irrelevant to learners in the 'Third World', importing Western curricula instead of being related to issues facing the local community and its socioeconomic environment. Despite these and other calls for environmental learning to be relevant, we know relatively little about learners' responses to environmental learning experiences in different contexts, and what the reasons for those responses are.

Nagel and Lidstone (2008) come to the rather worrying conclusion that for the students who participated in their study environmental education is at its best irrelevant and at its worst depressing. Battersby (1999) found that English schoolchildren's responses to environmental learning often involved judgements as to the relationship between what was being learned and their own interests, enjoyment, awareness, sense of agency and responsibility, and values. He argued that relevance is important in engendering positive feelings towards school and the curriculum. A similar conclusion was drawn by Smith-Sebasto and Walker (2005) in their research on learning experiences in a residential fieldwork context. These authors drew on the concept of relevance to learners to explain why students found some aspects of their environmental learning to be more meaningful and important than others.

Research also suggests that learners' engagement in and responses to learning experiences may also be mediated by judgements about its relevance to their own future (working) lives, however distant and imagined these may be. These issues have been highlighted with reference to students' experiences of school science (Driver et al. 1996) and geography (Biddulph and Adey 2004; Talbot 2002).

We use the term 'relevance to curricular context' to describe judgements that are made which involve the relationship between a particular learning experience (what is being learned and how: subject matter and task), and the setting of that learning (a particular course, subject, module, etc.). There is considerable evidence that learners' conceptions of subjects or disciplines are significant factors influencing their engagement in learning processes, and ultimately crucial mediators of learning itself (Buehl and Alexander 2005).

Brook et al. (1984) suggest that how a learner perceives the general curricular context of their learning (i.e. subject, discipline, course) will influence the way they contextualise, understand, and see the value in particular learning experiences. Research in science (Driver 1983; Driver et al. 1996; Shapiro 1994; Songer and Linn 1991) suggests that the way learners think about subjects can enable or constrain their learning. Broader conceptions of subjects or curricular units can underlie deeper learning processes, in particular the ways learners see different aspects of learning as linked together (Bruner 1960; Schoenfeld 2004). Several studies have suggested that when learners encounter environmental issues in school geography lessons they are likely to consider the relevance of particular content and experiences to the subject as well as to themselves (Dowgill 1998; Hopwood 2004, 2007a; Hopwood et al. 2005; Lam and Lai 2003; Norman and Harrison 2004).

As we discussed in the chapter titled "What Is Environmental Learning?" environmental education within formal settings is not typically constituted as a single subject in its own right. Instead, environmental learning may take place in a range of curricular contexts. This diversity of settings, combined with theoretical reasons and strong evidence that perceptions of relevance to curricular context do indeed mediate students' learning experiences, make a compelling case for seeking to better understand issues of relevance in environmental education.

The Importance of Differing Viewpoints Among Students and Teachers

At its root, exploring differing viewpoints amongst students and teachers is about (i) a dual focus on the teacher as well as the students and (ii) a particular interest in the curricular dimensions of teacher–student interactions/relations. The importance of examining these issues within environmental learning situations is threefold.

Firstly, looking at teachers and students together is important because it helps to avoid the tendency that has been seen in educational research in the past for work on teaching and work on learning to exist as separate fields. Marton (1994, p. 28), for example, spoke about 'the perplexing fact that two dynamic fields of research – teachers' pedagogical content-knowledge on the one hand, and students' understanding of the content on the other hand – are rarely connected'. A similar point is made by Cooper and McIntyre (1993, p. 383) whose work on secondary school classrooms was about bringing together what they described as the 'two established traditions of research on teachers' [thinking] and pupils' thinking'. Such studies have highlighted the complexity of student–teacher relationships and learning–teaching interfaces and the value that can stem from improved research-based understandings of these aspects of the classroom. This is particularly so for a topic like environmental learning which, as we have seen in earlier chapters, has rarely been the subject of in-depth research from the everyday teacher or, still less, everyday learner perspective.

The 'differing viewpoints' lens, however, is about more than a dual focus on teachers and students. The second reason for its importance lies in its very specific concern with the curricular dimensions of student–teacher interactions and relations. In the early 1990s it was recognised within the field of education that while there was a strong tradition of sociologically-informed work exploring students' reactions to the social aspects of schooling, such work has paid scant attention to the curriculum. As Erickson and Shultz (1992, p. 474) argued at the time, 'students are not seen as primarily engaging with and experiencing subject matter [but rather as] acting to create and negotiate the social relationships of which they are a part'. The limitation of this was that it overlooked a whole host of issues associated with the classroom curriculum such as how teacher–student interactions are mediated by the curriculum, how students engage with or disengage from subject matter and learning tasks, and the ways in which teachers' curricular decisions and strategies can facilitate and/or hinder student learning. If, as was argued in the chapter titled "What Is Environmental Learning?", environmental learning is about particular kinds of curriculum content, educational contexts, pedagogical strategies and educational purposes, then any overlooking of the curricular dimensions of educational settings will be decidedly unhelpful to efforts to understand and improve young people's environmental learning.

Finally, there is definite support for the idea of students and teachers being involved in 'negotiating viewpoints' from studies in other curriculum areas. A UK study of teaching and learning in secondary school history and English lessons, for example, described students' willingness to engage in lessons as 'a powerful negotiating tool' whereby 'students reward teachers who create opportunities for congenial classroom interaction with their engagement' (Cooper and McIntyre 1996, p. 94). Almost a decade earlier, studies of writing tasks in US high school English lessons revealed how some students were 'seemingly better able to model or interpret the teacher's conception of the task' (Nespor 1987, p. 221). The common point in both of these studies is not only that students are active agents in classroom settings, but also that the relations between teachers' and students' perspectives are a significant factor in the nature and quality of student engagement and learning. This is an important point because until now it does not seem to have been widely recognised or discussed by practitioners and researchers in environmental education. Apart from a limited number of examples (Schindler 1993; Elliot 1995; Cotton 2006b), there are few signs of environmental educators adopting a perspective or lens that takes into account students as well as teachers and looks specifically at learners' perspectives on the learning content and learning activities. This is problematic because it means that our current understandings fail to consider the ways in which environmental learning can present particular kinds of challenges for students.

Summary

Figure 1 provides an overview of the key questions and ideas associated with each of the three lenses. It also offers a schematic way of representing several of the key characteristics of the lenses as presented and applied in this book. They constitute different ways of understanding environmental education, but they do not provide the complete picture – either individually or in combination.

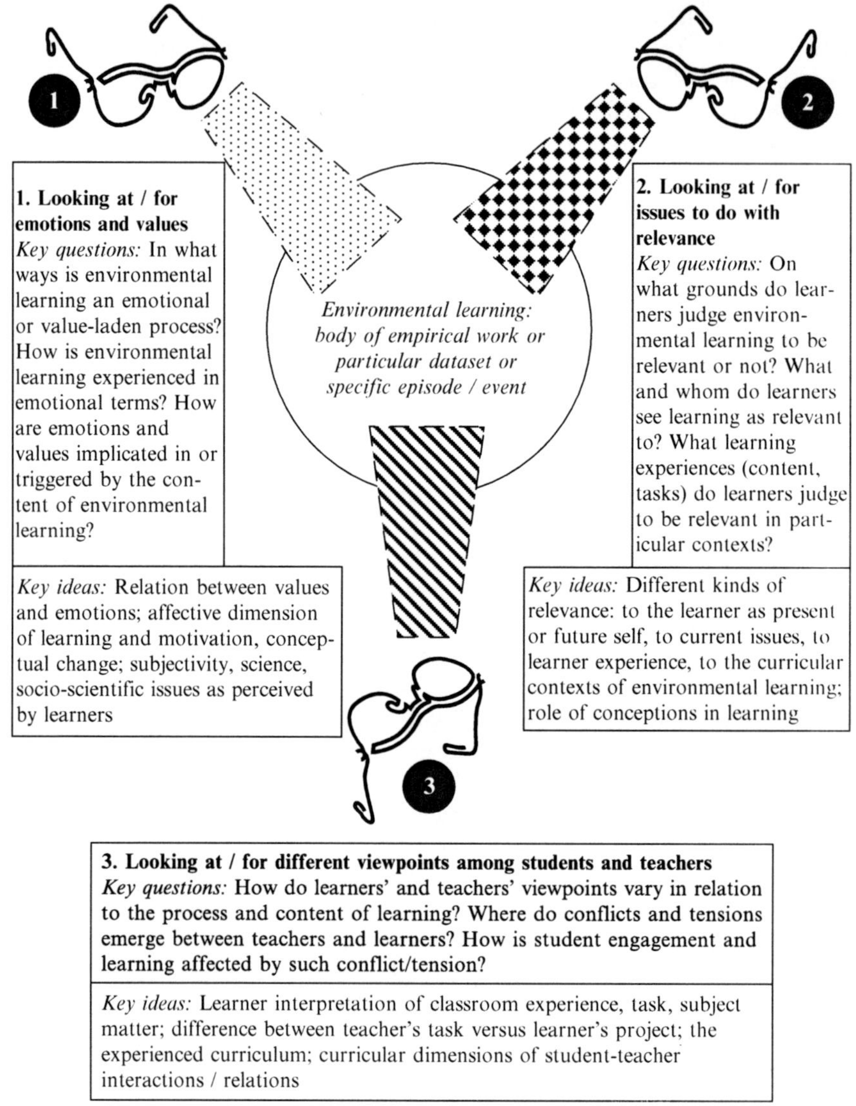

Fig. 1 Lenses for understanding environmental learning

The core of the diagram – the circle viewed through three lenses – indicates that they are different ways of looking at a common phenomenon – the phenomenon does not change, but what one asks about it, pays attention to, looks for, does change. The text within the central circle indicates how we see the lenses as means of looking at particular learning experiences or episodes, larger datasets, collections of empirical studies, or findings reported in the literature – all of which pertain to the broader phenomenon of environmental learning. The lenses have dotted borders because we do not see them as fixed or tightly bounded. Each has evolved through our own work, and we would encourage others to use them (or other lenses they construct) flexibly. Each lens is represented equally in size, but different in shading – depicting our view that none provides a more or less correct or truthful picture than the others – all offer equally valid, but distinct ways of seeking to understand environmental learning.

What is not represented in Fig. 1 (all such diagrams have a weakness associated with their simplicity) are the overlaps and relationships between lenses. For example, values and emotions are the primary focus of one lens, also emerge when looking at environmental learning through the others. This could be envisaged by swinging the spectacles round to overlap slightly.

Dealing with Emotions and Values

This chapter looks at environmental learning through a lens that focuses on students' values and emotions. We focus initially on the ways in which students deal with their own emotions and values when undertaking environmental learning activities. We then explore how students' perceptions of environmental subject matter as value-laden or otherwise can affect their learning. It should be added that while emotions and values are the primary focus of this chapter, they also emerge as part of later discussions concerning relevance (the chapter titled "Questioning Relevance") and student–teacher relations (the chapter titled "Negotiating Viewpoints Among Students and Teachers").

Introduction

In the pioneering work by Piaget and Vygotsky the emotional dimension was considered an important part of learning. However, research following these scholars has mainly come to focus on the cognitive dimension, at the expense of emotional and value-related dimensions (Jovchelovitch 2007).

Emotions and values are terms that encompass a range of different meanings, often referred to under a generic label of the affective domain. In this chapter, we explore students' engagement and disengagement and personal emotional responses or values stances in relation to what is being learned. Emotions can also be related to students' anxiety in school work and test situations, aggressive behaviour and self-control and metacognitive skills (Efklides and Volet 2005). Furthermore, the emotional dimension of learning is often connected with motivation and seen as an influence on students' engagement in learning activities. In recent years there has been increasing interest in the role of motivation, emotions and values in processes of conceptual development and change (Pintrich et al. 1993; Watts and Alsop 1997; Sinatra and Pintrich 2003; Sinatra 2005).

This chapter is structured in two main parts. The first examines students' emotions and values as part of the learning process. Various examples are used to illustrate the ways in which emotional responses feature within environmental learning.

M. Rickinson et al., *Environmental Learning: Insights from Research into the Student Experience*, DOI 10.1007/978-90-481-2956-0_5,

In the second part, we explore how students' perceptions of environmental subject matter can affect their learning. This focuses on the extent to which learners see environmental subject matter as value-laden and the ways in which they confront notions of subjectivity versus objectivity and facts versus opinions. In both cases we situate discussion of our findings in relation to wider literature. Given the focus of this lens, the literature to which we refer often takes us beyond writing on environmental learning per se and into realms of conceptual change, scientific literacy and socio-scientific issues.

Students' Emotions and Values as Part of the Learning Process

As discussed above, emotions may be seen as a driving force underpinning engagement in learning but they may also direct the focus of that engagement (Claxton 1989; Österlind 2006). Feelings of dislike or distaste may lead to total disengagement or partial disengagement in which students elect to focus on areas which avoid the need to confront or deal with issues they find uncomfortable. For example, Watts and Alsop's (1997) studies of physics undergraduates studying radioactivity show how emotions can exert an important influence on students' learning. This was seen in three ways:

1. Students who have difficulties in learning the topic but who make no reference to their like or dislike for the particular material
2. Students who are inhibited in their learning through distinct distaste for the topic itself
3. Those who have arrived at equilibrium between their wariness of the issues and an informed view of the risks involved (p. 359).

The authors stress the risks of disengagement in situations where students' emotional response is strongly framed by a feeling of unpleasantness or dislike towards the topic.

In line with the claims above, Claxton (1991) also argues that emotions are important for directing students' classroom learning, using the term 'stances' in describing students' different approaches to their work. They can, for example, focus on succeeding with exams and getting the teacher's approval (swot stance) but they can also strive for deep understanding (boffin stance). Claxton discusses this as being an unconscious process, where students are making a kind of 'cost–benefit-analysis':

> What they are 'up to' (in all senses) is the outcome of a tacit decision-making process based only on subjective estimates of competing priorities, opportunities, demands, resources and risks, which is, in their terms, sensible and vital. Only if the emotional/motivational factors in this decision permit or encourage intellectual learning and the subjective assessments are accurate, will achievement be limited by such cognitive factors as 'ability' or alternative conceptions (p. 159).

A wide body of work has thus pointed (empirically and theoretically) to the importance of emotions in the learning process. We continue this chapter by applying a lens focused explicitly on emotions and values to the specific context of environmental learning. We begin by contrasting two different situations from our research in which emotions emerged as part of the learning process. In the first, an emotional response leads to disengagement from the learning activity (Melanie in a geography lesson), while in the second, emotional responses divert students away from one focus and lead them to explore values-related issues more explicitly (biology undergraduates working on an open-ended task). This couplet of examples is followed by a more extended discussion of emotions and values in relation to learning about particular concepts or theories. Here we focus on undergraduate biologists studying the concept of price in economics.

Emotional Response to Content – Disengagement and Diversion

Our first example comes from one of the secondary school geography classrooms in Rickinson's study. Shortly after a lesson that involved watching two short pieces of film about rainforest destruction in Amazonia, Melanie (a 13 year old student) described how one of these films had provoked a strong emotional response for her. What is important is that her emotional response to the images within the film had a very real effect on her willingness to engage with the lesson (Box 1).

The case of Melanie can be seen as an example of a student pursuing a strategy of minimal engagement ('I was just paying a little bit of attention') as a result of emotional responses to the subject matter ('I don't like cutting trees down, I don't like animals being hurt or moved'). Of course environmental subject matter does not necessarily provoke such responses but given the controversial and emotive nature of much of the subject matter implied by environmental learning, we might expect such encounters to be far from uncommon.

Box 1 Melanie's disengagement as a result of emotional responses to the subject matter

The second video ... didn't seem very interesting – cutting down trees, I don't think that's very my sort of thing, I don't like things like that. I don't like cutting down trees, I don't like animals being hurt or moved or anything. I don't like anything like that. Yeah I'm very against it, I think it's awful.

I was hardly watching it, I was just paying a little bit of attention 'cos I don't watch things that are boring. If I think they're boring I don't watch them ... I just judge things before I see them and I didn't think it was really very exciting ... If I've got my own opinion on them, yeah I like hearing other people's opinions because it's their opinion, but no matter what you can't change my opinion. You can try till you're blue in the face but I will not change.

Emotions played out differently among a group of Swedish biology undergraduate students working on solving an open-ended task on environmental audit reports produced by companies. Here the intended focus (as expressed by the teacher and understood by the students) was to analyse what constitutes a company environmental audit report and how one might evaluate such a publication. However, the students' understanding of this task was affected by their emotional responses in terms of their negative feelings towards the companies and their environmental credibility. They found themselves repeatedly talking about and analysing the companies rather than their environmental audit reports.

Hans, one of the four students, reminded the rest of the group of the focus they had decided upon initially – to study the reports and not the companies. In the excerpt provided in Box 2, which comes from the group's last discussion about this task, Hans emphasises the need to focus on analysing the reports. The discussion centres on the issues that provoked the strong feelings, and thus the students' engagement may be interpreted as being directed or perhaps diverted by their emotional responses.

This example contrasts with that of Melanie – here emotional responses do not lead to disengagement, but rather engagement in a related discussion which is itself shaped by the feelings experienced by the students. In a similar way to Melanie, though, we can see the way in which students' feeling and emotions about the topic (namely the environmental practices of companies) influenced their engagement with the learning activity. While there are differences between the minimal engagement of Melanie and the diverted engagement of the biology students, what they have in common is students having to deal with their own emotional responses to environmental subject matter.

Students' Values in the Process of Understanding Concepts and Theories

Another way in which emotions and values can feature in environmental learning is in terms of students' responses to particular concepts or theories. We saw this with undergraduate biologists encountering concepts like 'price' when studying economics (Lundholm 2007, 2008b). Before we look at this example, though, it is helpful to briefly highlight recent literature emphasising the role of emotions in conceptual understanding.

In the book *Intentional Conceptual Change*, Sinatra and Pintrich (2003) argue that more needs to be understood about the internal as well as social and contextual aspects of conceptual change and learning. Southerland and Sinatra (2003) discuss these issues in the context of learning about evolution:

> Evolution is a complex topic that is inherently difficult to learn even when ones personal beliefs do not conflict with the content. The situation becomes even more complex when firmly held religious beliefs are perceived to be in direct conflict with the scientific explanations of human evolution in general, and heightened for explanations of human origins[…]

Box 2 Biology students grappling with their feelings about companies

Having worked over a period of 3 weeks, the question of whether the companies are actually making any environmental commitment or progress keeps coming back.
Lena: But if we formulate a question, like this: Do environmental audit reports describe reality at all? We can't judge that, if it's a mirror of reality since we don't know what reality is.
Karin: No, we don't know the companies well enough.
Lena: But we can speculate, sort of. We can't answer the question, but we can pose the question.
Nina: Okay, then I also think we can discuss [in the students' written report] that an important aspect of good environmental work within the companies, **is** that the staff is involved in the environmental work, and not only the boss –
Lena: But I've written that already.
Nina: Yes, but Hans questions that.
Hans: Yes, I don't think it's relevant to the focus of our work, which we decided upon from the beginning.
Nina: But you haven't reacted to that in our discussion.
Hans: No, but I admit, it's really difficult. I have to think all the time: What am I looking for, really? It's like thinking in math about logarithms; every time you see them you have to think, what does this mean?
Karin: Yes, it's hard to stay on track.
Lena and Nina: It's very difficult.
Karin: We can't get too emotionally involved, that's the problem!
Hans: No.
Karin: We can't actually include what we *think* of them.
Nina: But yes, we can include what we *think* of them.
Karin: No, not what we think about these companies and their –
Lena: We can't say "I think they're crap!"
Nina: No, exactly "X is a hell of a company, cutting down the forest" – no!
Hans: We share a common view of the companies –
(All the students are talking at the same time. Difficult to hear the different voices.)
Nina: Yes, yes, but I think the same way – companies suck! (She laughs). *That* judgement we should not include.
Karin and Lena: No.

> Thus, whether intentional constructs come into play when learning new scientific concepts may depend on the complexity of the topic, the degree to which the content is perceived to be in opposition to the learner's prior conceptions, and the perceived conflict with the learner's broad, culturally-based belief systems. (pp. 336–337)

These authors conclude that when students see the topic as controversial then factors such as learners' personal epistemological beliefs and readiness to question their own beliefs and values become particularly important. Previous studies have suggested that students may retreat from economics study because of resistance to the idea of monetary value. This is despite recognition among learners that 'such approaches can serve to achieve objectives consistent with their original beliefs (such as wildlife preservation)' (Shanahan and Meyer 2006, p. 104). The example below of biology undergraduates grappling with the concept of price in economics illustrates these issues clearly. Of particular note is the way in which the notion of price was considered by the students in the context of closely-held values relating to nature.

Two students, Anna and Diana, were giving accounts of their views of the subject of economics. Anna found pricing itself problematic as 'everything has to be shown in dollars and cents when a decision is to be made, and my world view really opposes that'. Yet, she is eager to learn and tries to combine her biological knowledge with economics. Diana also described the clash with the subject in terms of what is considered as 'the base' in the subjects of biology and economics respectively. Encountering economics meant 'flipping the pyramid' and considering nature as an 'appendage' rather than the foundation for economic activities. The quotes below express these experiences in the students' own words.

> You have to think in terms of money all the time. Not values or something like that, but the fact that everything has to be shown in dollars and cents when a decision is to be made. My worldview really opposes that. You have to price everything because the world is governed by economy, sadly enough, so you have to price it otherwise these economists won't understand. It's difficult for me, because often enough the value you put on nature is minimalistic, the real values are so much higher, and so often they are valued less than they should be. (Anna)

> The difference between us biologists and the economists, is that they often think there is a 'base' in the economic world, or possibly in the social world, while we see the economic and social world depending on ecology as a foundation. So, it's the opposite really, the world upside down… The way I see it, nature constitutes the base, we are part of nature and through nature we get economy, that means, what is traded is taken from nature, it's not something that's just there like magic in a factory, it has its origin in nature. I think that's why I found it very difficult in economics; they picked out the environment part, and sort of put it aside, like an appendage, instead of the way I see it; as the foundation for everything. If nature wasn't there we wouldn't be here ourselves. (Diana)

We can see how these students' perceptions of the discipline's overarching worldview, particularly as it relates to nature, presented them with genuine difficulties. As they related pricing of goods and services to nature (air and ocean) and services such as ecosystems services (biodiversity and pollination), they found the subject increasingly problematic. This is illustrated further in Box 3.

We may speculate that in these examples values act both against and for the learners. On the one hand, values (as well as knowledge) of nature create obstacles in the students' coming to understand key concepts and theories in the discipline. On the other hand, it is precisely these values that have led these students to choose a course in economics and to gain knowledge that they see as useful and important in their future profession: making the world environmentally sound.

Our findings illustrate in the context of environmental learning a phenomenon that has emerged in recent studies in other contexts. Sinatra (2005) discusses the

Box 3 Examples of students' emotions and values in learning about the concept of price

Cornelia
These big ecosystems services from the ocean and the air, nobody owns that. The rights of owning are perhaps limited to, what's it called, a fishing certificate, and there's no ownership, no selling and buying, so how to find a value?

Diana
I don't think everything should be given monetary value, it's actually impossible to do I think. It's more of a thought experiment, it might even be dangerous to do, since you have to take time and change into account. It can be terribly misleading. But also, it's very individual. That is, what value you give, it depends who you ask.

Fanny
Personally, I don't think you can use economics all the way, because the world is much too complicated, for example the natural resources will come to an end – there's a big conflict. I mean, how is that dealt with in economics?

'warming trend' in conceptual change research, and our research has indeed suggested the importance of values and emotions not only in motivational terms (underpinning or directing engagement in learning, as discussed above) but also in the 'nitty gritty' of learning about particular concepts. The key point here is that students' difficulties with environmental learning activities can be as much due to issues of emotions and values as to challenges of knowledge and understanding.

Students' Conceptions of Values in Subject and Subject Matter

While the discussion thus far has focused on emotional responses to subject matter, this section highlights a different way in which emotions and values can feature in environmental learning. Here the focus is on how students may perceive environmental subject matter as neutral, objective, factual (on the one hand) or values-rich, subjective, and opinions-based (on the other hand). Our findings show how students' responses to learning activities can be related to values they perceive in the subject matter and to their own perspectives on such values. We find students raising questions like: How do we distinguish our own viewpoints and values from factual knowledge? Are opinions different from knowledge?

Questions such as these can be seen in relation to work on epistemological beliefs and learning (Mason 2003). In this respect notions of scientific literacy and socio-scientific issues are particularly relevant. While the aims of environmental learning may well be much broader than being scientifically literate (as discussed in the chapter titled "What Is Environmental Learning?"), there are some important parallels and overlaps between scientific literacy and the capacity to interpret evidence and argument in relation to complex environmental issues. This section therefore begins with a brief discussion of some key ideas relating to scientific literacy, before moving on to consider the ways in which students' perceptions of values in environmental subject matter played out in our studies.

Scientific Literacy

Scientific literacy has been proposed as an important aspect of what students need to know not only to understand science but also to be able to participate in public debates and make justifiable decisions on scientific – often environmental – issues presented in the media (e.g. Driver et al. 1996). Although the meaning of the term varies, researchers interested in this area have stressed a number of common features such as critical thinking, understanding the nature of science and epistemological awareness. In particular there is a frequent emphasis on understanding how science and scientific knowledge develop.

Scientific literacy also entails skills of argumentation, the understanding of arguments and rebuttals and how these are justified and can be challenged. In relation to this issue there has been an interest in how students use scientific knowledge in decision-making on socio-scientific issues. Grace and Ratcliffe (2002) explored students' understanding of conservation management and their study revealed that when students made decisions, their arguments were seldom based upon scientific knowledge. Instead, values and ethical considerations were noticeable. This connects strongly with some of the students in our research, where (as discussed below) there were clear examples of learners trying to navigate between science, political viewpoints, economic interests and ethical concerns.

It has also been suggested that scientific literacy should include students being able to make moral judgements through moral reasoning. Several contributors to *The Role of Moral Reasoning on Socio-Scientific Issues in Science Education* argue that moral understanding is important in order to make sense of socio-scientific issues such as environmental problems (Zeidler 2003). Two threads in the literature on scientific literacy can be identified which are of particular relevance here. The first relates to issues surrounding the role of opinion and emotion in what is often conceived as a value- and emotion-free scientific process, and by extension to learner perceptions of subject matter and the nature of knowledge. The second points to the fact that more recent writing on science education has foregrounded the importance of learning about complex issues that have both social and scientific dimensions (often rooted in some kind of environmental problem).

Values in Environmental Subject Matter

Drawing on these ideas concerning scientific literacy, we can now consider another important dimension highlighted by our 'emotions and values' lens – students' perceptions of values in environmental subject matter and in particular the role of facts and opinions and objectivity and subjectivity. Our first example returns to the Swedish undergraduate biologists mentioned earlier who were analysing environmental audit reports as part of a course entitled 'Environmental control for biologists'. This is followed with several examples of English secondary school students' experiences of environmental learning in geography lessons.

In their work on environmental audit reports produced by companies, four of the students came to pose the following question: Is it scientifically correct to be subjective? The group discussed in what ways the kind of task and work they were engaged in was different from previous studies in the field of natural science (Box 4). One of the students, Nina, answered the question by stating that the group had tackled the issue of subjectivity since they had all scrutinised and analysed the environmental audit reports using the same pre-agreed criteria. The group's discussion on this topic is concluded by Hans and Nina describing what they think should be included in the report.

In the excerpt below Hans turned to Lena and said she would prefer writing a column, as in a newspaper. By using the word column, Hans highlights the difference between scientific writing and other genres. Consequently, he implicitly clarifies that personal views and opinions do not fit and are not valid in the academic setting.

Issues relating to facts, opinions and personal viewpoints also emerged in Hopwood's study of secondary school geography. Here a similar discourse is apparent, framed in terms of being 'right' or 'wrong' in contrast to having opinions or views (paralleling the notions of objectivity and subjectivity which framed the university students' work). Box 5 documents an interaction between the researcher (Hopwood) and a student (Ryan) which took place after several lessons in which students were explicitly asked by their teacher to make personal choices and explain them. Here it is clear that Ryan perceives geographical knowledge not in terms of ultimate 'right' or 'wrong' qualities, but rather as a reflection of different viewpoints or 'sides' in arguments. He feels that there is scope for students to come to their own conclusions or express their opinions, and that what appears to be the most appropriate response is not a matter of 'right' and 'wrong' but more a matter of what perspective one takes. Ryan thus perceives geographical subject matter as something about which he may have his own opinion, but also something about which several different views may exist. He described how he relates his opinions to the knowledge and perspectives given: 'It's just there to help give you a better idea of what your opinions are… It puts some ideas into your brain and helps you'.

Two further examples illustrate the variation in learner perceptions of values and subject matter in school geography. Commenting on his learning during a sequence of lessons about hurricanes, Matt stated that there is a difference between facts and

Box 4 Biology students debating the role of values in scientific writing

Hans: They didn't say it should be a scientific report, but it should be of 'scientific appearance' and there are certain frameworks; you state what kind of material you've used, how you worked things through, and present it logically, perhaps not chronologically, but in some kind of order any way and –
Nina: I think it sounds great!
Hans: – and then you 'blur' in the discussion.
Nina: Well, material and method – this is what we have used as a method; we have checked the companies according to specific criteria and these are the results; we have seen and found this and this, and, yes, then there is a discussion, okay, – how have we interpreted this?
Hans: Lena wants to write a column or something. Do you agree Lena, is this the way to go?
Lena: No (laughs), well, yes, that'll be okay.
Hans: So, more writings about the basis for these criteria ... And then write that we were four, that we read different reports, so that we describe the way we conducted this work. There's the risk of some arbitrariness.
Nina: Exactly. That's something we should stress.
Lena: Really? It isn't that bloody scientific –
(All students talk at once. Inaudible)
Hans: But we should explain the way we worked, shouldn't we? If we had done a lab report or the like, then we would have been very thorough.
Lena. Yes, but then you're detailed in every aspect, but that's not the case –
Hans: And then I think that we should describe, that, concerning 'results', there's a certain arbitrariness.
Lena: Meaning that all we've written isn't really true! (laughs).
Hans: Noo, but –
Lena: I was just joking!
Hans: But one shouldn't hide the fact that there's subjectivity –
Lena: No.
Hans: ... in this, that can differ within the group as well.
Lena: I'm very subjective! Hurmph! (Clears her throat)
Hans: Yees! (everybody laughs). In some way or another we should describe the way in which we've conducted our work, the division of the material – we don't have to write who read what.

opinions within the subject itself in that descriptions of hurricanes are 'facts' whereas the effects of hurricanes are a matter of opinion. Matt perceives descriptions of the natural world (phenomena, processes, features) as factual (right or wrong, true or false) and descriptions of human life (impacts of natural hazards, culture, lifestyles) as grounded in opinion rather than fact, and therefore not subject to right/wrong criteria. The quote below illustrates this point in his words:

Box 5 Ryan's view on the role of personal opinions in geography

Interviewer: Should geographers be looking at both sides in all things they do?
Ryan: Yeah definitely, in most things. It depends, because even though they have got their own opinions, they should still look at both sides of the argument and then give a conclusion on their own opinion.
Interviewer: In choosing an answer, how would someone think that's a good or bad choice, how would you judge them?
Ryan: If someone said to me "I'd chop it all down because of the money and everyone would be happy making all the money" I think they're only looking at the people who want to make money like the hotel owners that need all the furniture [i.e. from the trees that are cut down] rather than the people in poverty that don't have anything, probably what we have, so I'd probably say to them have you thought about those people in poverty? …At the end of the day I think it's all about your personal opinion and what it comes down to is your personal opinion.
Interviewer: So in geography there's no particular view, it's up to the individual?
Ryan: No it's not all up to the individual in geography. It's just there to help give you a better idea of what your opinions are, so if you said that the whole of Brazil was rich, geography will come in and say "it's rich in some places but – this, and this, and this" and it puts some ideas into your brain and helps you.

> You should really know the facts and then how it [a hurricane] was formed was basically core stuff, really basic stuff. But then how it affects people's lives; people have many opinions on that, so you write whatever you really want to write on that.

Yet another view on opinions, values and subject matter was explained by Jenie, who claimed that school geography is all about opinions, 'an opinion subject' in which your personal opinions don't fall into right or wrong categories.

> I don't think there is a right answer, I think it's more a kind of opinion subject because people have different opinions about things, you can't be right about something or wrong, I think, because it's your opinion.

We can explore these issues further by considering the responses of classmates Ryan and Jenie to a series of lessons comprising tasks explicitly focused on exploring different points of view in relation to deforestation in tropical rainforest ecosystems.

Ryan and Jenie both describe the way different views are an important part of the subject (geography) and how it has developed their thinking of environmental issues. For Ryan, it means learning about 'the bigger picture', and coming to an understanding of deforestation from a contrasting view than his own. Jenie also talked about the importance of giving voice to different views but in light of a social aspect suggesting that it is important that everyone is heard and acknowledged, and stressing it is important 'not be selfish'.

> Ryan: I used to think that the rainforest was chopped down not for reasons, just because they chopped it down to sell it, I found out why, what advantages it's chopped down for, and the disadvantages of it not being chopped down, and I always used to think that there was only one reason.
>
> Interviewer: How has this made you better at geography?
>
> Ryan: Well I think you improve so much each lesson. I have enjoyed it, I have become a better geographer because I've learnt how to look at a bigger picture. Before I only looked at the picture, which was what I believed in, which was against deforestation and now we've just done that I can see more of a bigger picture and more reasons why some people think that.
>
> Jenie: I think people have their own view and they don't look at anyone else's, but when you do a debate, it kind of makes you stop and think about other people as well, so you're not just 'this is what it should be, blah, blah'. Like the counter-argument thing, you have to think about other people and what they would say.
>
> Interviewer: Do you think a good geographer should do that?
>
> Jenie: I do, because I think you shouldn't be selfish and think about your own view, you need to think about other people as well.

Students' notions that geography is an 'opinion subject', appear to resonate with earlier findings on students' conceptions of the social sciences (Molander 1997). These examples illustrate a series of issues which emerged across the data from all three studies. Across contexts – from undergraduate biologists to school geographers – students were grappling with their perceptions of the nature of environmental subject matter. Perceptions varied within contexts (classmates and group work peers often had different views) and were associated with varying views as to the nature of particular knowledge domains, the appropriateness of particular kinds of thinking, argumentation and writing, and the relationship between particular curricular contexts and environmental issues.

Values in Solutions to Environmental Problems

We have discussed how students may have varied views of environmental subject matter (i.e. the content of learning) in relation to values, opinions and facts. A slightly different, but related issue emerged in Lundholm's study of engineering undergraduates' experiences of an ecology course. Here perceptions that environmental problems are value-laden and politicised led students to consider the role of values and opinions in considering solutions to those problems (a focus on trying to find solutions being particularly important in the context of learning to be an engineer).

One of the students, Ola, having talked about contrasting views on nuclear power, finished by exclaiming: 'There are a lot of values in a subject like this. I mean, just think about environmental issues!'. Another student, Patrick, stated that solutions to the question of society's energy supply are political matters.

> When it comes to solutions, it's all about politics. I mean nuclear power, it's all politics and talking: shutting them down and buying Danish coal power instead, what's the advantage of that? ... The solutions that exist today are probably stemming from politics on these matters.

The ways these undergraduate students are reasoning on 'subjectivity' and 'opinion' in relation to environmental topics resonate with findings of Zeidler et al.'s (2002) study. They investigated students' conceptions of the nature of science and reactions to evidence that challenged their beliefs about socio-scientific issues. The authors conclude that 'students distinguished between the 'subjectiveness' of opinion and the 'objectivity' of scientific knowledge' (p. 352). Students associated scientific knowledge with proven, tested or constructed knowledge, whereas subjectiveness was seen as personal opinions. Another interesting finding is that students completely separated science knowledge from personal knowledge and opinion, arguing that the former had little influence on the latter.

> Although certain students viewed scientific knowledge as that which is supported by concrete evidence and facts, they would not consider the use of scientific evidence to convince other people to change their personal opinions. It would appear that these students felt that opinions are immune to change despite any concrete evidence provided that supports alternative viewpoints. (p. 360)

The Swedish students cited above (Ola and Patrick) seem to be distinguishing science from opinions along similar lines to the students in Zeidler et al.'s study. However in the case of our examples, these perceptions were extended into students' consideration of how solutions to environmental problems might be tackled.

Box 6 Undergraduate engineers struggling with a lack of discussion about environmental solutions

Interviewer: How does this political aspect influence the discussions in the classroom?
Patrick: We haven't had time to discuss, unfortunately.
Interviewer: I've noticed there hasn't been any discussion.
Patrick: There's no time for that.
Interviewer: Would you want more time?
Patrick: Both yes and no. I mean, of course it has to be discussed, I mean if you don't discuss it here, where else should you discuss it? It should be addressed here, at a higher level than for example the level in media ... I think it's of importance to ventilate and discuss solutions [to environmental problems], and perhaps what ideas we have and what kind of new technology there is, where the possibilities are.
Tobias: It feels like there's no time for debating these matters, instead we should just do this and this, and, this is wrong, and this. You don't have time to talk about it! I think it's weird.
Interviewer: There hasn't been any discussion ...?

(continued)

Box 6 (continued)

Tobias: No. That's it, how can we improve things? It's all 'this is wrong, and this'... But that's in general as well, there's too little time for everything ... It all seems very condensed and there's no time for debates, it's all about: "this is what you have to know! Get it in your heads best possible way".

Our final examples come from students expressing concerns about not having discussed solutions to environmental problems (Box 6). Patrick argues that discussion informed by appropriate knowledge is what is needed as opposed to what he sees as more superficial media debates lacking in knowledge and in-depth analysis. For Tobias, it is more of a problem of insufficient attention being given to debating possible solutions.

These examples show how students' perceptions of the nature of the issues they are learning about (complex, controversial, liable to values-based solutions) combine with their personal values as engineers to frame their responses to and interpretations of learning experiences focused on findings solutions to environmental problems.

Summary

In the chapter titled "What Is Environmental Learning?" we suggested that much of the subject matter at the heart of environmental learning may be characterised as controversial and contested. In this chapter we have examined our findings through a lens focused on values and emotions to probe further into how learning is experienced by – how it feels to – learners. As will be the case in the next two chapters, the key points from this chapter are now summarised with reference to the who, what, where, how and why framework of environmental learning outlined earlier (Fig. 1 of the chapter titled "What Is Environmental Learning?").

Much of the discussion in this chapter has focused on the relationship between the 'who' (i.e. learners) and the 'what' (i.e. subject matter) of environmental learning. Furthermore, throughout this chapter we have presented examples of learners not only learning about nature and society (and relationships between them), but simultaneously confronting, articulating and (re)considering their own views. In this sense, this chapter has incorporated all elements of the model we used earlier to conceptualise the what of environmental learning (Fig. 2 of the chapter titled "What Is Environmental Learning?") but has continually emphasised the learners themselves (their values and emotions) as closely interwoven with subject matter.

We have seen examples of strong emotional responses provoked by environmental subject matter, responses which may lead to minimal engagement in learning activities (e.g. Melanie and the rainforest videos) or to highly engaged discussions which divert from an intended focus (e.g. Lena, Karin, Nina and Hans and the environmental audit reports task). Emotions and values also emerged in the process of learning about specific concepts as with Anna, Diana, Cornelia and Fanny's struggles with price and nature in economics.

The weave between learners and subject matter was also seen in learners' perceptions of the nature of the subject matter they encountered. A key issue here was the extent to which students saw environmental subject matter as either neutral, objective and factual or values-rich, subjective and opinion-based. This question played out amongst biology undergraduates dealing with the role of values in scientific writing, school students expressing different views of the role of opinions and facts in geography and undergraduate engineers thinking about environmental solutions. In all of these contexts issues relating to the 'why' of environmental learning became more apparent – students held various conceptions of environmental learning as learning received 'factual' knowledge, learning about others' views and/or learning about one's own views. Here we also found students dealing with issues relating to a more activist or professional sense of environmental learning – learning to find solutions to environmental problems.

Questioning Relevance

This chapter focuses on the second of the three lenses we have used to look across the data generated in our three studies. Here we look at or for issues to do with relevance, and consider how these are manifest in environmental learning processes from the perspective of learners. We explore a range of environmental learning experiences and learners' comments on these experiences. This analysis pursues two related themes: the first explores how environmental learning may (or may not) be conceived as relevant to learners themselves, while the second considers learners' perceptions of the relevance of environmental learning to particular curricular contexts. Finally we reflect on what this lens shows us, linking back to the issues and themes discussed in the chapter titled "What Is Environmental Learning?", and outlining key points which are developed further in the chapter titled "Enhancing Environmental Learning".

Introduction

Relevance is often taken for granted as a valued and aimed-for characteristic of learning. However we lack a clear understanding of what issues of relevance mean to learners, and how these relate to their experience of learning. This is particularly the case with respect to research in environmental education. Our understanding in this regard has been expanded and refined as we have shared data from our three studies. We describe this process of looking across research findings with a specific focus as involving the application of an analytic *lens* which guided our interpretation of evidence as we looked at or for issues to do with relevance. In the chapter titled "Lenses for Understanding Environmental Learning" we explained why relevance is important, highlighting how research and theory (within and beyond environmental education) suggest that learners' perceptions of the relevance of what they are learning act as important mediators of learning experiences.

As indicated in the chapter titled "Lenses for Understanding Environmental Learning" (see Fig. 1 in that chapter), this particular lens frames analysis in terms of questions such as: On what grounds do learners judge environmental learning to be

M. Rickinson et al., *Environmental Learning: Insights from Research into the Student Experience*, DOI 10.1007/978-90-481-2956-0_6,

relevant or not? What and whom do learners see learning as relevant to? What learning experiences (content, tasks, activities) do learners judge to be relevant in particular formal contexts? What are the grounds upon which learners make judgements about the relevance of environmental learning? What learning contexts do students see as suitable for exploring environmental issues? Our exploration of these issues is underpinned by a number of ideas which have been developed and discussed in wider theoretical and empirical domains outside of environmental education. These include distinct notions of relevance (to present or future self, to curricular context) and the role of conceptions and perceptions in learning. Learners' ideas of relevance appear to have a strong bearing on their engagement in learning processes and their conceptual notions of relationships between the various things they learn.

This chapter does not provide a conclusive or exhaustive response to the questions posed above (the empirical contexts are limited to formal settings and do not encompass the full range of curricular contexts for environmental learning). Nonetheless we offer an enriched understanding of how learners may construe the relevance (or otherwise) of environmental learning.

Relevance to Learners

Introduction

In this section we examine environmental learning through a lens focused not on subject matter per se but on learners themselves and how they relate to that subject matter. When considering this – the relevance of environmental learning to learners – we explore two different kinds of relationship. The first remains in the moment, and involves asking if and how learners see environmental learning experiences as relevant to themselves in that particular place and at that particular time – in the here and now of the learning experience. The second asks questions about how knowledge or learning activities may be seen as relevant to the learner as they envision themselves in possible future scenarios. This analysis of relevance to the learner has parallels with the notions of self-oriented subjective assessments of learners discussed by Randler et al. (2005), but refracts a broader notion of relevance in terms of learner interests in related but distinct ideas.

Relevance to Self in the Present

Educators in many contexts (across disciplines and educational phases) are frequently urged to ensure that the learning experiences they offer their students are relevant to those students and their everyday lives. Evidence suggests this is important because learners appear to be more motivated and engaged, and to perceive greater value in learning that they see has some relation to themselves as individuals and their personal experiences. While there is some empirical evidence addressing these issues in the context of environmental education, our understanding is far from complete.

The arguments and examples we provide below suggest perhaps a more complex picture than might be expected, one in which individual agendas, emotions and values are significant. We see many parallels between the issues discussed here and those presented in the chapter titled "Dealing with Emotions and Values". We begin our discussion by considering three learners, all studying school geography (see Table 1), and then move beyond these to explore similar issues in relation to undergraduate engineers.

Jenie often commented on her learning experiences in terms of their relevance to herself. In her case the invoked self was very much in the particular place at that particular time. To have this kind of relevance, in Jenie's view, learning should relate to the places and issues she is familiar with, has experience of, and feels engaged in. This can be expressed as relevance contingent upon the coincidence of learning with personal experience.

In one lesson Jenie and her classmates wrote down a series of questions which framed their subsequent investigations and fact-finding about Brazil. Afterwards she indicated her frustration with these experiences, strongly articulating how she deemed them to be irrelevant to her: 'In lessons like that it just doesn't mean anything to me because I've never been there before. If it was somewhere I'd been on holiday or something I could relate to it more... it completely means nothing to me'. Here Jenie contrasts the place she was learning about with the places she is familiar with. Significantly she does not suggest that to be relevant learning must focus on her immediate environment, but rather explains that if she is to feel some kind of connection with learning about different places and environment.

A similar issue emerged in relation to a debate about rainforests. She explained that she did not care about the rainforest in the Amazon, but added 'if there was a rainforest on the [school] field it would be different'. Thus it was not the notion of rainforests themselves that Jenie thought was irrelevant, but the sense of spatial separation and lack of relevance to the kinds of environments and surroundings with which she is familiar.

Interestingly, Jenie found lessons about sustainable development to be more relevant to her. After participating in a role-play which reproduced global trading

Table 1 Examples illustrating notions of relevance to present self

Example	Environmental learning is most relevant to me when
Jenie	... it is about places where I live or where I have been, issues I am involved in (learning coincides with personal experience)
Sara	... it involves studying people, environments and places of which I have little knowledge or experience myself (learning about the unfamiliar)
Lisa	... it addresses issues of human impacts on the natural environment, something I feel strongly about (learning about issues I care about)

Source: Hopwood study.

inequalities, she commented 'it was more relevant because it [global trade] had a bigger effect and more of an effect on me.. it sounds really selfish but it is more interesting to learn about something that affects me personally'. She explained how the personal shopping choices she made were part of the global trading system they had been studying, how she could support sustainable development by buying fair trade products.

We have also described Jenie's notion of relevance as grounded in a sense of the present. Over several lessons pupils in her class had explored issues such as the links between deforestation, carbon dioxide emissions and global climate change. When commenting on the sequence of work, she remarked 'we don't do about things that have affected us'. While she saw a global scale impact resulting from deforestation of tropical rainforests as likely and indeed important, she doubted whether any effects would be noticed in her lifetimc, and on this basis questioned the relevance of her learning.

Jenie's sense of relevance *to me here and now* resonates with the ideas put forward in the literature relating to environmental education and other contexts (Battersby 1999; Brooks 2003; Connell et al. 1999; Lynch 2003). What we find interesting are the rationales and criteria she applies, for example why she found issues of fair trading sustainable development to be relevant, but the deforestation of tropical rainforests irrelevant to her. The links learners draw between subject matter and their own lives and experiences are not necessarily straightforward or as might be expected.

When Sara considers the relationship between environmental learning and herself, like Jenie she thinks of her own location, context, knowledge and experience. However she argues it is more relevant (and necessary) to learn about things she doesn't encounter in her everyday life or know about already: we express this as seeing learning as relevant when it relates to the unfamiliar. She sees such learning as a vehicle to expand her horizons, to go beyond her present context and location.

After watching a video about the impact of hurricane Jeanne on people in Haiti, Sara commented 'it was good learning, because I'm not really going to go there but at least I know about it'. After the following lesson, on the impact of hurricane Gloria in the United States of America, she remarked 'you need to know about other people as well as yourself, that's important innit, because the world, you should know about it'. In other instances she explained how she thought it was important for people like her to learn about different cultures and places around the world. All these examples involve learning experiences centred around environments, people and places that were at a distance from Sara, and which lay beyond her own life experiences. She did not see these as irrelevant, but rather relevant to her as a learner because without studying them she would not know about them.

The contrast between Jenie and Sara is clear. We do not see cases such as Sara as mounting a direct challenge to those who encourage teaching and learning about environmental issues that can be related to learners' own experiences, contexts and everyday lives. However, it does suggest that there are no guarantees that learning about the familiar will be perceived as relevant to learners. We would interpret Sara's position as placing value on (school) learning in terms of escaping the familiar, fitting her values which assert the importance of learning about other people and their relationship with the physical environment. This interpretation is supported by a wealth of evidence from the original study in which Sara consistently demonstrated concern for other people's wellbeing, particularly in the face of extreme environmental conditions or events. Thus we present Sara's notion of relevance as relevant to her, in the here and now, as a young person who cares about other people and who sees learning about (environmental) issues beyond her immediate surrounds as important.

With Lisa we find a further example of notions of relevance to the learner here and now, only in this case issues of familiarity or unfamiliarity were less central, and instead her comments foregrounded the relationship between learning and a general set of environmental issues. These issues were important on a personal level, relating on occasion (but not always) to her personal experiences, and were reflected in aspects of her outward identity. The following quote encapsulates aspects of her sentiment and personal experience of human encroachment on what she saw as natural areas: 'I'm a 'farmy' girl. Where I ride [horses] is a farm. I used to go at this place in Springfield but all the fields have been turned into houses, so I go to this other farm place now'.

Lisa cares passionately about the (physical) environment, or in her words 'nature'. She describes herself as an 'ecofriendly kind of girl' and uses a Greenpeace pencil case, as well as adorning her school bag with badges and patches that testify to her 'green' identity. Talking more generally about school geography, she describes a green subject that 'is about protecting nature' and in doing so 'making the world better'. Her judgements as to the relevance of particular learning experiences involved considerations of relevance to curricular context (i.e. school geography – discussed below) and considerations of relevance to herself as an 'ecofriendly' girl. She felt learning about population structure and fertility in the United Kingdom and Tanzania was relevant to her so that she 'personally can do something about it' – 'it' being the negative effects that population growth has on the environment, her personal role being to avoid having too many children and encourage people around her to do the same. Similarly she felt that learning about tropical rainforests was relevant to her so she could 'realise how important they are and if you cut them down it's gone for ever'; as a result she would try to avoid personal implication in deforestation through not buying products made of Amazonian wood.

The examples of learning contexts in the previous paragraph include aspects both close to Lisa (the United Kingdom, her sense of her own fertility) and more distant from or unfamiliar to her (Tanzania, tropical rainforests). Lisa's focus on human–nature relationships illustrates one example of a more generic quality: judging relevance to the self in terms with reference to a set of issues that are important to the learner (in terms of identity, interest, ability to act, desire for change, etc.). Evidence of a similar nature from the context of Swedish engineering undergraduates is outlined in Box 1. Here the issues to which learners feel a personal connection and thus relevance share a common but broad association with climate change and/or the greenhouse effect. In these cases the learners' personal commitment to environmental issues is also translated into a shared sense that learning is not only relevant to them, but also is relevant to society as a whole, related to what can be described as a collective socio-moral imperative.

Evidence of similar issues emerges in research from other contexts too. Smith-Sebasto and Walker (2005) report different foci of the reflection undertaken by American 10–14 year-old schoolchildren participating in a formal environmental residential programme. These included notions of relation to the personal self, particular interests and emotional responses. However there were also examples of perhaps an even more imminently 'relevant to me here and now' notions of relevance in terms of immediate safety in an unfamiliar and potentially hostile physical environment, such that safety sessions were deemed 'more relevant than academic or recreational sessions'. Although somewhat departing from the classroom-based contexts that we focus on in this book, this illustrates the incidence of similar phenomena to those highlighted in our own research in a wider range of settings for environmental learning.

Box 1 Relevance of environmental learning to Patrik, Sara and Jenny (Lundholm study)

These students' comments illustrate how learners may perceive relevance to themselves and to society more widely.

Patrik discussed learning about environmental issues such as the greenhouse effect: 'I mean things that everyone ought to know… common knowledge.. basic facts to understand what they see on TV. It's even in the book for driver's license about the greenhouse effect, so it's really basic, things you should know in society today. It shouldn't have to be in the driver's license book'.

Sara links the idea of relevance to the private self with a wider 'everyone should': 'I don't just think it's for work, instead it's more private sort of, you get a reminder about the environmental problems ... I almost think everyone should take the course'.

Similarly Jenny stressed professional value alongside personal relevance which is situated in a more general sense of what people should learn: 'I think it's important not just for work, but it's important to get some *allmänbildning** about the environment and ecology. Actually I think you have use for everything, no matter what kind of profession you'll have in the future'.

*A Swedish word meaning general educational fostering; Schnack (2008) discusses related ideas of *bildung* and *allgemeinbildung* from the Danish tradition in relation to health and environmental education.

Relevance to Self in the Future

We now turn our attention to a second set of ways in which environmental learning is deemed relevant (or not) by learners, to a sense of future self. We explore the case of Ryan (Hopwood study) who sees aspects of environmental learning in school geography as helping him develop skills (practical and conceptual) that will be useful to him in the future. We also present and discuss comments made by Swedish undergraduate engineers studying an ecology module (Lundholm study) who have a specific and defined sense of future self (as engineers) which is used as a benchmark for assessing relevance.

After a series of lessons on sustainable development, Ryan spoke about his experiences and ways in which he thought they were relevant to him. He explained how he had developed skills that would be relevant to his future working life through his learning on this topic:

> It's developing our future. She [the geography teacher] is helping everyone and it's a way to develop everything, so research when we're older. Because we're all going to be doing jobs, if you're doing engineering you're going to be developing and achieving, so like research and helping everyone do that.

In relation to studying banana cultivation he described learning (relevant) skills of 'trying to get information about how we're developing and information to help us', and in the context of studying rainforests commented he had learned to 'research what had to be done to keep things sustainable'. These examples illustrate clearly the relationship between classroom environmental learning experiences and the learning of research skills which, in Ryan's eyes, would be relevant to him in later life. Similar notions of relevance focused on practical skills are reflected in Smith-Sebasto and Walker's (2005) research in which American primary schoolchildren commented on learning something meaningful and relevant to them because they had learned a new skill.

Ryan also discussed another set of skills that he thought constituted features of environmental learning that were relevant to a projected future self. These are more akin to thinking or conceptual rather than research or practical skills. In this case the future self was less separated from the present self, and we interpret it as a sense of continuity from present into future contexts. After studying ecotourism and sustainable management strategies for tropical rainforests he was asked whether he thought his learning was relevant in any way. His response encapsulated an idea that was repeated in other learning contexts, particularly those focused on relationships between people, the physical environment, and development. He said that these learning experiences had made him 'a better person, more [able] to look at a different side of lifestyles rather than keep looking at yours. It helps everyone and makes you more understanding'.

On other occasions he spoke of learning to 'go into other people's shoes, to hear other people's opinions'. Thus he saw his learning as relevant to him, not in terms of the facts or content so much as in terms of the skills involved in understanding and taking into account a wide range of perspectives. This has an evident quality of relevance to him in the here and now (consistent with the examples relating to other learners given in the previous section), but there is also a strong suggestion of projection into the future. Ryan often spoke of environmental issues in terms of 'future living and changes' or 'what we're going to come against in the following years', and being able to tell future generations 'what issues they could face and what they could do about them'. Thus when we interpret Ryan's comments through a lens focused on relevance, we are inclined to interpret a sense of enduring relevance in the sense that the learning of skills he sees as relevant to him now is carried forward in his reasoning to situations in which he might find himself in the future.

We turn now to focus on the Swedish undergraduate engineers taking an ecology course. Martin explained how he took a consciously limited approach to learning when he felt the content was less relevant to what he imagined being useful or needed as a practising engineer. He thought detailed study of the water, nitrogen and phosphorous cycles and associated chemical formulas were irrelevant, and that his learning on this topic would soon be forgotten: 'I can still tell you the cycle of nitrogen but in a couple of months I won't [be able to do so]. But I don't know if it's important to know, what shall I use it for? In that case it is perhaps more important to know that it is not renewable'. The final words of this statement signal the shallow limits of what he deemed worth learning and thus how his notions of relevance (to him as an engineer) influenced the way he learned and what he learned. The lack of relation to a projected future self is particularly strong in the following comments, which this time relate to learning about the water cycle:

> Ecology, I don't know, let's say the water cycle. You don't really need the big picture. It's enough to know that in this place water runs from here and on ... and then the fact that it doesn't relate to what you want or to what you are here for. It is kind of on the side so to say. So it becomes, it feels, I don't think you have to take the ecology course to become a good engineer. It is only a thing you have to do to get through.

We can see similar judgements according to a future professional self in the comments of Jenny and Sara outlined in Box 1: both refer to relevance as 'not just for work' – implicitly implying that at least part of the relevance of this environmental learning was to do with their likely practice as engineers. In the case of applied, vocational or more explicitly work-oriented courses such as engineering or teacher education, the distinction between notions of relevance to curricular context and relevance to future professional self is perhaps less clear-cut, because the present curriculum is seen to be preparation for that future self. In an example from a different study Julie, an American university student training to become a school science teacher (Brandt 2007) complained about the 'lack of science content' in a landscape-focused aspect of her course. Traces of both relevance to curricular context and an invoked future self (in this case as schoolteacher) can be seen here. This multiplicity of interpretation is consistent with our notion of lenses – they are not mutually exclusive, and within each there are more subtle distinctions which may provide alternative ways of looking at a particular phenomenon, finding, or dataset.

Summary

We have begun our exploration of relevance by focusing on examples collected around a common theme of relevant to the present self, to 'me, here and now'. We have seen that some learners, like Jenie, deem learning as relevant when it relates to their own personal experience, context and location, when learning coincides with what is familiar to them. For others, like Sara, the opposite is true, and relevant learning is described as that which provides the learner access to the unfamiliar – to people, environments, and issues that are at a remove from their own lives. Here we might stress the idea of 'relevance to me as a learner'. This contrasts learners like Lisa who discuss relevance more akin to 'me as a committed activist' in the sense that relevant learning is that which relates to environmental issues that the learner cares strongly about both in the formal learning setting and in their personal interests, lifestyles, hobbies and identity. We have also pointed to ways in which learners may consider the relevance of learning in relation to their future selves – both in more vocationally-oriented curricular settings (undergraduate engineers), and in the classroom context of a conventional subject discipline (school geography).

Relevance to Curricular Context

In this section we explore learners' ideas about how (if at all) environmental learning is relevant to particular curricular contexts. By curricular context we denote a range of formally bounded learning contexts: school subjects, modules, courses, programmes of study at university. We discuss variation in students' beliefs in this respect, and the significance of these beliefs in influencing the experience of environmental learning.

Environmental education is rarely constituted as its own curricular unit per se. Thus when learners encounter environmental learning experiences it is often in the context of a separate (but perhaps related) curricular context (Connell et al. 1999). Learners' ideas about what constitutes this space can be important in influencing their response to and engagement in learning experiences that have an environmental quality or character to them. We were struck by Littledyke's (2004) finding that 'few [pupils] had a perspective of science as an important influence in society and as a factor in environmental issues, either as a contributing element to the problems or as part of the solution' (p. 230). In contrast to this apparent ruling out, Connell et al. (1999) report a ruling in, quoting one student who felt it was 'not possible to learn anything about the environment unless you took particular subjects' (p. 102).

Learners may have quite specific ideas about subjects or other curricular units, and whether these incorporate environmental issues or not. The purpose here is to dig deeper into these issues and elucidate what they may mean for environmental learning.

Looking across our three studies we found that students tended to discuss relevance to curricular contexts with reference to physical and human phenomena.

They did not always use these terms, but we have found them useful in making sense of learners' ideas, and the distinction, conceived in broad terms, also appears applicable to evidence from other studies reported in the environmental education literature more generally. Other terms used by participants in our research include nature, science, environment (physical), and man-made, people, built, culture (human). These terms have different meanings but for the purposes of our discussion the physical/human labels are sufficient to convey a degree of shared meaning on the one hand (grouping related ideas together), and a mode of distinction on the other (differentiating ideas from each other). While serious challenges have been raised in relation to such dualisms (e.g. Whatmore 1999, 2003), these themes authentically reflect the ways in which participants in our study experienced and discussed their environmental learning.

Relevant to Learn About Physical or Human Phenomena

As we explained in the chapter titled "What Is Environmental Learning?", we share with many other authors a notion of environmental learning as *not* limited to learning about the physical or natural environment. However it became clear from our own research and studies reported elsewhere that learners' notions of the subjects or programmes they are studying are not necessarily so broad.

The boundaries learners draw around particular curricular contexts may lead to exclusions of certain areas of content. We found evidence that some students consider either physical or human aspects to be relevant, but not both. These ideas may be highly significant in influencing the process of environmental learning. Learners may, for example, choose to ignore aspects that they believe fall outside the boundaries associated with a particular subject, module or course. Or, they may experience negative affect because they feel they are not making progress that is relevant to the goals of that particular context. Such judgements of relevance may also influence the ways students conceptualise subjects or courses as a whole and link different aspects of their learning together. We begin our exploration of these issues with an in-depth discussion of one learner, Matt, affording attention to detail in a specific context. After this, we demonstrate how similar issues emerged in a range of contexts.

The Case of Matt

Matt (Hopwood study) clearly articulated his idea that in school geography the main focus is on scientific or factual explanations of physical processes. He used the terms 'environment' or 'nature' to denote things like weather, climate, the sea, hurricanes, earthquakes, volcanoes. These he contrasted with 'culture', 'lifestyles' and 'feelings' which he argued were not really geography. After lessons about hurricanes, rainfall, and global climatic variation, Matt consistently explained how he saw the physical aspects – scientific understandings of natural processes – as those that were most relevant to geography. When discussing a series of lessons on hurricanes, he commented:

> The main thing would probably be how it [a hurricane] had formed, what path it takes, why it takes that path ... I mean more natural things ... I think how it is formed *is* geography.

Observation notes and documentary evidence clearly show how many of the learning experiences discussed with Matt were explicitly set up by the geography teacher as involving issues relating to people. The same lessons were experienced very differently by his classmate Sara, who considered both physical and human phenomena relevant to the subject. The following paragraphs explore how his ideas are associated with a filtering of focus and engagement in learning processes, notions of quality in work he produces, and his broader conceptualisation of geography as a subject.

Matt's engagement in learning tasks and activities was focused on those aspects he saw as most relevant to the subject. Thus, in a wide-ranging and open-ended extended writing task at the end of the hurricane topic, Matt produced a report that gave detailed explanations of physical processes (how hurricanes form, where they occur, etc.), and was replete with facts and figures such as the assigned category of hurricane, top windspeeds, required sea temperatures for hurricane formation and so on. He used a map to support his explanation of how hurricanes form (with arrows indicating their origins in the warm equatorial seas), rather than (as was the case with his classmate Sara, discussed below) depicting the hurricane's path across land and inhabited areas. His report paid relatively little attention to human aspects, save for quoting measures of destruction as indicators of physical force and power. His ideas of what is relevant to geography mapped on to what he envisaged being relevant in other assessment situations (academic relevance). He commented after a lesson about how people cope with a hurricane as 'not so important because you don't necessarily need to learn about it for a test'. His work from this lesson contrasts the highly detailed diagrams and explanations he produced in a subsequent lesson on what causes rainfall.

Matt's view that relevance to geography means a focus on physical aspects also appears to be central in his more abstract understanding of the subject and in the concepts he uses to link different aspects of his learning together. He explained his idea that 'geography is all about the world', and that various aspects of geography are related because 'they are all to do with the world'. He uses the term 'the world' to denote a physical entity (i.e. earth), referring to materials such as rocks, land, and water, and processes such as wind and volcanic eruptions. He constructed a concept map showing relationships between a range of topics he had studied in geography, and placed at 'the world' at the centre. Everything was linked to this central idea, whether through the notion of material origin in the earth or physical changes due to global processes (e.g. seasons affecting trees).

We find much of interest and indeed a challenging nature in this example. It strikes us that Matt's environmental learning experiences are subject to a strong filtering-out of many aspects to do with people, culture, feelings, etc. Evidence of selective (dis)engagement in particular aspects of learning activities, and physically focused notions of quality, envisaged assessment, and more abstract (subject-wide) conceptualisation poses a significant challenge and may well be to the detriment of Matt's learning. We can imagine arguments in favour of breadth (attending to human and physical aspects, especially in interaction) as having some value over the increased depth and detail within only one strand that Matt sees as most relevant. Further his developing understanding of environmental features, processes and issues might be enhanced by a less exclusive sense of relevance and a greater openness to considering human aspects.

Other Learners and Contexts

The detailed discussion of Matt illustrates issues that we found in several contexts, across our three studies and in the wider research literature. In Box 2 we present a second example from school geography in the United Kingdom, and in this case we find evidence of negative affect in Aiden's response to classroom experiences deemed irrelevant to the curricular context. We find very similar issues arising in the context of American university students training to become school science teachers (Brandt 2007):

> Another student, Julie, who was completing a Masters programme to be certified as a high school science teacher was extremely frustrated by the sociological emphasis of the class [which was focused on landscape], and she complained bitterly about the 'lack of science content' in the course. The student was firmly invested in the norms of Western science ... (p. 13).

Box 2 Aiden thinks learning about people in the rainforest is not relevant; Jo disagrees

Aiden (a participant in Rickinson's study) discussed relevance to geography as a school subject in ways that were very similar to Matt (see main text). His boundaries of relevance *included* physical features and processes, and *excluded* aspects that were more focused on people. This was particularly apparent in his response to a lesson about Kayapo Indians, inhabitants of the Amazon rainforest. He commented that the focus that day had been more relevant to history than to geography. More precisely, he explained that (in his view) while the rainforest itself is relevant to geography, the people living there are not. These comments were associated with a strongly negative assessment of the lesson overall and different parts of it (including a video about Kayapo lifestyles). Other pupils in Aiden's class felt differently: Jo found the strange practices of the Kayapo intriguing, and said the focus on people was both relevant to geography and made 'a nice change'. This indicates, as in the case of Matt, that notions of what counts as relevant to particular curricular units may be highly individualised.

Here we see boundaries of relevance constructed around a 'science' (relevant) versus 'sociology' (not relevant) dualism that maps broadly onto the physical and human concepts framing our discussion. The transgressing of these boundaries in a particular learning experience resulted, as with Aiden, in a strong sense of negative affect. Other literature suggests these issues and ideas may be relatively common in the context of school science. Littledyke (2004) aimed to explore how young people perceive science in relation to their learning about environmental issues, and found evidence that the two were often seen as unrelated. Sauvé (2005) notes that the link between environmental education and science teaching has been much debated from the perspective of curricular design and instructional leadership, offering the following as a comment not atypical of many voices in these discussions:

> Environmental education threatens the integrity of the disciplines. We risk draining science of its disciplinary content. If, for instance, we introduce environmental issues, we are not doing chemistry. Values education is not science! (p. 18).

We also found instances of similar responses to digression from what is perceived as relevant to a course in the context of Swedish undergraduates taking an ecology course as part of a degree in engineering (Lundholm study). These further examples fit the pattern of relevance constructed around a particular curricular unit, again mapping broadly onto a physical/human dualism. Jenny and Karin had discussed with each other their thoughts that the ecology course was poorly adapted for their civil engineering programme. They found the focus on the physical environment rather irrelevant, and would have seen learning about built environments and their impacts on ecology as more appropriate to their course of study.

Tobias found the ecology course to emphasise physical systems and problems rather than considering adaptations and solutions. He contrasted this with the approach more typical of his engineering studies more generally, which focused on thinking of solutions to problems, creating man-made ways to work around the problems imposed by the physical environment. After the course he had a general impression of ecology as 'just so much mumbo-jumbo', lacking relevance to the applied realm of engineering.

Unlike the examples of Matt, Aiden (Box 2) and Julie (Brandt 2007), these learners *included* human aspects as relevant to, and *excluded* physical aspects from the curricular unit they were studying. These examples demonstrate further the negative affect and self-limited or managed approaches to learning that may occur when learning experiences are perceived to lie outside what is deemed relevant in a specific context.

Relevant to Learn About Physical and Human Phenomena

In this section we explore evidence that unlike the learners discussed above, others may deem both physical and human phenomena to be relevant aspects of environmental learning in particular curricular contexts. We show that students may draw boundaries around subjects or courses that are inclusive of both dimensions rather than exclusive of one or the other. Furthermore, we indicate evidence suggesting how such conceptions of relevance may relate to subject-wide conceptions and understandings of links between different aspects of learning. Physical and human aspects may be seen as two distinct but relevant strands, or as more integrated in relation to a particular curricular context.

Physical and Human as Relevant but Separate

Jenie (Hopwood study) used a range of terms to describe geography as a school subject. These consistently built on the idea of a dual-stranded subject with two distinct aspects, sometimes referred to as 'the people side and the non-people side'. According to Jenie the former involves learning about people's lifestyles, religion, cultures, feelings, and points of view; the latter was described as 'trees and natural stuff, like natural unmoving things that don't have legs. Volcanoes and things'. These were however deemed by her to be separate foci in the subject: 'I think there's too halves to geography. There's like the man made bit and then there's the natural bit'. Unlike Matt (discussed above) Jenie did not disengage from aspects of learning about either the people or non-people sides (to use her words) because she felt both were relevant. This translated to an approach to learning which often (but not always) focused on only one at a particular time. Further details of the evidence elucidating her ideas are provided in Box 3.

Box 3 Jenie describes both physical and human aspects as relevant, but separate

Several interactions between Jenie and the researcher demonstrated her ideas about the inclusion but separation of physical and human aspects in the context of school geography. When considering a series of images, she classified them as relevant to geography, and in her explanations either situated them as relevant to a people-focused strand or a nature/physical-focused strand. Similarly she responded to a series of questions, arguing they were relevant either because they related to people or to the natural environment. Jenie was asked to take photographs showing what geography was about, and she later explained how her set of images show variation between physical and human phenomena studied in geography, although each image alone was designed to depict only one of these. In a concept map (in which she linked various aspects of geography together), Jenie described relationships within these two 'sides' rather than across them, referring to shared characteristics within physical or human categories.

Jenie's approach to environmental learning in this context was often associated with one strand or the other at any particular time. For example, in one lesson pupils were asked to list a series of questions they would like to explore in relation to Brazil. Jenie wrote in her book 'Population. How urban is it? MEDC or LEDC [More/Less Economically Developed Country]? How many people are homeless? What are the main cities?' These questions then provided the frame for her subsequent investigation in class. After the lesson she talked about her work and explained that her questions were all focused on 'the people side' but that questions about the physical environment (about rainforests, climate, rivers) could also have been relevant.

In a different lesson Jenie was asked to consider different options for tourism development in a rainforest area. Her written work focused strongly on benefits of tourism to people. In this case she commented 'I'm not looking at it from a tree's point of view, 'I'm a tree, I'm going to get cut down', it's more about people, if it's going to affect people'. This focus was maintained in an extended assessment task about which she noted 'mine is more about how people were affected by [activity in the rainforest]'. Evidence shows these selective foci were not a reflection of an exclusive sense that people were relevant and the physical environment irrelevant. Rather she saw both as potentially relevant, but chose to focus on one of the two at a time.

Physical and Human as Relevant in Interaction

We can contrast the example of Jenie with those of Lisa and Sara (Hopwood study). All three girls said human and physical phenomena are relevant to geography. While Jenie conceived them as separate foci within the subject, Lisa and Sara saw them as overlapping and intertwined. These ideas were manifest in their engagement in particular learning activities, as well as in their conceptions of how different aspects of their learning linked together. We now consider each of these learners in turn.

In Lisa's case there was a strong, direct link between geography and the idea of learning about the environment: she characterised the subject as about studying relationships between people and the environment (readers will notice the strong resonance between her view of geography and her general values and interests). Thus her ideas of relevance included both physical and human aspects, in interaction. She saw different aspects of geography as linked together through their sharing a common feature of addressing issues of how people affect nature (as with pollution) or how nature affects people (as with natural hazards). Box 4 describes these ideas and evidence for them in more detail, and may be compared directly with previous examples relating to Jenie.

Lisa's ideas about geography, and particularly what she thought was relevant to geography, were powerful influences on her engagement in and response to environmental learning. We can consider a series of lessons focused on technical

Box 4 Lisa describes both physical and human aspects as relevant, in interaction

Lisa was given a disposable camera and asked to take photographs which represented her ideas about geography. Her six images depicted what she described as variation in the impacts people had on nature, including a car (illustrating pollution and damage to the planet), a built up area, a building site for an airport ('having an effect on where it was built because all that would have been fields'), a mixed grassy and residential view, and a series of fields (which Lisa saw as entirely 'natural'). Lisa referred to people as well as built environments when discussing 'human' aspects, for example stating that 'anything to do with urban areas' would be relevant to geography because of the pollution associated with conurbations and their constituting a man-made imposition on a natural landscape. The concept map Lisa produced to show how different aspects of geography link together was structured around a series of relationships between physical and human phenomena. For example she linked climate to three more people-related aspects: tourism, development, and population.

procedures used to construct population pyramids in order to represent population structure, and on learning about the different factors which affect fertility and mortality rates. These experiences were introduced by the teacher as part of a broader topic on population, and were explicitly couched in terms of human processes (such as infant mortality due to lack of healthcare, high fertility as a result of particular religious beliefs, delayed marriage and birth patterns due to shifting working lives, etc.). On the surface, as presented in the public classroom discourse, these experiences did not seem directly related to geography as described by Lisa: a subject centred on relationships between people and the environment. Analysis of Lisa's written work revealed little in terms of how she made sense of this experience in the broader curricular context as she conceived it.

However interviews conducted after each lesson showed that she did indeed perceive them as relevant to geography and associated environmental issues. This is apparent in the following interaction:

> Lisa: When you think about it logically you do get out how the environment is going to change through population.
>
> Researcher: What do you mean?
>
> Lisa: Say you are looking at a certain environment, you need to know what the amount of population is going to do to that place, like [in] the countryside there's a lower population and more fields, but in urban areas there is a lot more people and buildings.

A similar logic was apparent in Lisa's discussion of lessons about migration, and the factors which push people away from a particular place and pull them towards a different place. She commented 'it's about population and going back to how it affects the world'. Lisa's engagement with and perceptions of learning about global climates and small scale weather patterns followed a similar pattern.

We now consider Sara who, like Lisa, believes both physical and human aspects are relevant to the geography as a school subject, studied in interaction rather than as separate strands. In her view, geography is different from other subjects because of its focus on both people and the physical environment as they vary across space. She contrasted religious studies, in which 'you learn a lot about people' with geography, where 'you learn that bit more about why people are because of where they are or where they live'. This builds on her idea of a close interrelationship between human and physical phenomena in a particular place. For Sara learning about physical environment is not separate from learning about people; the two are closely intertwined.

Sara's engagement in classroom tasks appears to have been influenced by these beliefs. Unlike her classmate Matt (discussed above) her work focused not on physical processes per se, but on how physical processes are related to human experience. Sara's report on hurricane Ivan explored physical processes and their impacts on people, as the excerpt in Box 5 shows. The general approach is one which stresses the significance of physical processes in the context of human lives, focusing on interactions between the weather system and people (rather than just

Box 5 Excerpts from Sara's report about hurricane Ivan

Sara's written work clearly reflected her focus on both human and physical aspects in interaction. The diagram she used to illustrate her explanation of physical processes included a representation of a city, and the annotation 'the hurricane will move towards the city'. The sentences below are drawn directly from her work and illustrate how it was consistent with her notion of what counts as relevant in a geographical context.

Hurricanes destroyed towns and cities, villages and most areas. It causes houses to flood, destruction of buildings and even death. Hurricane Ivan was no different! Hurricane Ivan was started near Grenada and finished just as it hit Cuba. Its power destroyed houses, buildings and caused deaths. Millions of remaining buildings were left without power, trees were ripped up from the earth and key buildings were destroyed including hospitals! 90% of Grenada was left in devastation.

Hurricanes kill and completely ruined cities, towns and villages. However there are a few safety precautions that might save some lives. People are urged to stay nearer the ground perhaps in an underground basement if possible ... Florida is perhaps more prepared for a hurricane than some of the islands. Perhaps this is because it has more money to build places of safety and their buildings are made from stronger material. Little islands are often very poor and cannot afford many safety precautions! This causes a severe amount of deaths and islands can be completely destructed.

the physical processes themselves, as was characteristic of Matt's approach). Her writing distinguishes the experience of the hurricane in Florida from that in the Caribbean islands. She paints a picture of a tight interweave, with people being subjected to physical processes, and at the same time those processes varying in impact according to the human context in which they occur.

The contrast between Matt and Sara was maintained in subsequent tasks relating to processes of rainfall formation, which were similarly construed by Sara in terms of physical–human interactions while Matt's work focused on physical processes. Sara introduced ideas of spatial variation in order to incorporate a dimension of human implications in her work. Both students produced diagrams and explanations relating to three different types of rainfall. Sara's were distinctive in that both visually and in the text she made reference to the everyday experience of people, their livelihoods, cultures, and the threats that weather conditions may pose. Similar contrasts were evident in these two pupils' work on tornadoes: Matt concentrated on scientific explanations of physical processes, Sara constructed a narrative of what it would feel like to be in a Tornado to accompany her writing about how they occur.

Summary

Looking across our studies and the wider literature we have found that learners may have strong beliefs about what counts as relevant to particular school subjects or university courses. These beliefs may vary within a particular group of learners, may not always be explicit, and may be at variance with the views of teachers, curriculum planners or others with a vested interest in environmental learning.

We have focused on concepts of the human and the physical because these emerged as strong themes in our data. They enable us to illustrate the ways in which learners may deem environmental learning to be relevant to particular curricular contexts or not, but they are by no means the only concepts which are relevant (Bart, for example, in Hopwood's study, deemed anything relating to the future irrelevant to geography).

Perceptions of relevance to curricular context are not just important because they influence the learner motivation and judgements about the value of a particular task or activity in relation to assessment and achievement in that context. These ideas of relevance can influence the way learners approach and engage in tasks, whether associated with a focus on areas deemed more relevant and disengagement from other areas, or when seen as part of the general understanding of curricular context that learners use to link different aspects of their learning together. We suggest that further discussions around what counts as relevant to particular contexts, or around where environmental learning takes place, should pay close attention to the learner perspective, and avoid assumptions that views held by teachers and other educators outside of the learning context will be shared by those engaged in the learning process itself.

Summary

As in the chapter titled "Dealing with Emotions and Values", we can reflect on what looking at or for issues of relevance tells us about the experience of environmental learning by drawing on aspects of the framework we presented in the chapter titled "What Is Environmental Learning?" (Fig. 1 of the same chapter). This framework draws attention to who is learning, where they are learning, what they are learning, how they are learning, and why they are learning.

This chapter first explored the theme of relevance to learners, demonstrating the importance of understanding links between the *what* and the *who* of environmental learning. Across our studies we found students making judgements about classroom experiences according to how they related to themselves as individual learners. These judgements were not merely *post-facto* evaluations, but in-the-moment responses and reactions which mediated engagement in classroom tasks and activities. In our discussion we distinguished between relevance to self in the present and relevance to self in the future, providing examples of how both of these may be invoked in learners' accounts of their experiences. When we think about who is engaged in environmental learning, we must not only think of learners as school students or university students, but also learners as they envisage themselves in the future. This reminds us also to consider *why* students are learning, what motivates and drives them to engage, and confirms the importance of including learners themselves in conceptualising what students learn about in environmental education (see Fig. 2 of the chapter titled "What Is Environmental Learning?").

Learners may experience complex, multi-layered reactions to environmental subject matter. They may feel removed and distant from the issues at hand, but at the same time feel it is personally important for them to learn about them (as with Sara). They may find aspects of content unappealing, but recognise their potential importance in their imagined future professional roles. They may identify strongly with particular agendas, and interpret or reframe learning experiences so that they fit the mould which they find most interesting and important (as with Lisa). Lack of relevance to imagined future roles may emerge alongside a perceived social–moral imperative to learn and understand certain environmental issues (as with Patrik, Sara and Jenny). Rather than thinking of subject matter as relevant or not, it makes more sense to think of the varied grounds for judgement and points of reference that may be brought to bear in learners' perceptions of relevance. Even when the point of reference is the self, there may be more than one way in which perceptions of relevance are played out.

The second theme explored in this chapter focused on relevance to curricular context – addressing important issues relating to *where* environmental learning takes place. Environmental education rarely constitutes its own curricular space, and thus we were interested in exploring learners' perceptions of the relationship between environmental learning and particular subjects or courses. Our findings show that learners may hold strong beliefs about what counts as relevant learning depending on the curricular context. This is not simply a question of mapping different aspects of environmental learning into different subject domains (such as geography, science, ecology, etc.). Learners differ significantly from each other in their views, and even within a single classroom, different opinions may be held and judgements made as to what aspects of an environmental issue are relevant and should be brought to bear when working in a particular subject. We presented examples illustrating how learners may ‘rule in’ or ‘rule out’ different aspects, and follow their own personal logics of relevance through so that they influence what they choose to do, what they envisage achieving, and how they choose to do it.

There are many calls within and beyond environmental education to make learning relevant, and many good reasons why researchers and teachers should focus their energies on pursuing this goal. Examining empirical findings through a lens which focuses on relevance demonstrates that relevance cannot be taken for granted, and is certainly not for curriculum planners alone to decide or control. Looking at or for issues of relevance as they emerge in students’ accounts of environmental learning has shown us that perceptions of relevance do indeed matter, but that the relationships between environmental subject matter, curricular context, and learners (with their varied interests, values and projected future selves) are complex. Relevance is thus a rather slippery and elusive target, but nonetheless one which we must engage and explore critically if we are to understand the experience of environmental learning.

Negotiating Viewpoints Among Students and Teachers

The preceding two chapters have demonstrated the potential of exploring environmental learning situations using lenses focused on the role of emotions and values and issues of relevance. In this chapter, the analysis continues using a lens that includes the perspective of teachers as well that of learners. Our focus here is on environmental learning situations in terms of the interactions between the viewpoints of students and the viewpoints of teachers. Using this lens, we consider four types of teacher–student conflict situations that arose within our studies of school and university environmental learning. These concern differences in viewpoints between teachers and students on: environmental issues; what is topical or controversial; what is relevant for a specific curriculum subject and the nature and value of empathy tasks. We finish by drawing out some key messages concerning the what, where and how of environmental learning.

Introduction

This chapter examines environmental learning situations in terms of the interactions between the beliefs and values of students and the beliefs and values of teachers. The lens here draws attention to the ways in which differences between teachers' and students' viewpoints can present challenges for student engagement and learning. As argued in the chapter titled "Lenses for Understanding Environmental Learning", the idea that classroom teaching and learning can be influenced by differences between students' and teachers' views is highlighted by research in a number of curriculum areas. Studies of writing tasks in US high school English lessons have revealed how some students are 'seemingly better able to model or interpret the teacher's conception of the task' (Nespor 1987, p. 221). Meanwhile, UK research on teaching and learning in secondary school history and English lessons described students' willingness to engage in lessons as 'a powerful negotiating tool' whereby 'students reward teachers who create opportunities for congenial classroom interaction with their engagement' (Cooper and McIntyre 1996, p. 94).

M. Rickinson et al., *Environmental Learning: Insights from Research into the Student Experience*, DOI 10.1007/978-90-481-2956-0_7,

Both of these studies highlight ways in which the relations between teachers' and students' perspectives can affect the nature and quality of student engagement and learning.

Within the field of environmental education, there seems to have been little interest in exploring the dynamics of student–teacher interactions and relations in real-life learning settings. There are few signs of environmental educators adopting a perspective or lens that takes into account students as well as teachers and looks specifically at learners' perspectives on the learning content and learning activities. One important exception is the OECD Environment Schools Initiative (ENSI) Project which involved participating schools collecting data on their teaching and learning processes through interviewing students and teachers (Elliot 1995; OECD 1995). In the cross-national report on this project, Elliot (1995) describes how the Danish teacher participants experienced problems in developing work on domestic animals because they held completely different understandings of domestic animals from their students. As these teachers explained:

> We had agreed [with the pupils] to work with domestic animals but we had obviously different concepts of domestic animals. We had meant production animals and the pupils had meant pet animals [...]. They preferred to work with domestic animals, and therefore, they were surprised when we introduced the work on farm animals [...] It was no wonder the atmosphere was a bit listless. (Elliot 1995, p. 34)

Along similar lines, one of the Austrian ENSI teachers described how his attempt to develop independent environmental project work with a class of 13 year old female students led to a major student–teacher conflict situation. Like the Danish example, what became clear through further reflection and sharing of experiences within the class, were significant unstated misunderstandings between the teacher and the students. One issue, for example, was that the students 'saw the project as the teacher's business and not their project' (Schindler 1993, p. 457). Both of these examples from the ENSI Project show differences in viewpoint between teachers and students having a significant impact on how students feel and act in relation to environmental learning activities.

Within our studies, there were several cases of students talking about their environmental learning activities in terms of some kind of conflict or tension between themselves and the teacher. What is important, though, is that these were not general interpersonal conflicts or tensions. Instead, as with the Danish and Austrian examples above, they related specifically to values and beliefs connected with the environmental content/learning activity. More specifically, they concerned students and teachers having differing viewpoints on:

- Environmental issues – where students perceive there to be a significant difference between their own perspectives on an environmental issue and those of their teacher.
- What is controversial – where students and teachers held contrasting views on what is controversial.
- What is relevant – where students and teachers had different views on what is relevant within a particular curriculum subject.

- Empathy tasks – where students and teachers think quite differently about learning activities designed to develop empathy.

Differing Views of Environmental Issues

One area of difficulty for students within our studies was where they perceived there to be a significant difference between their own perspectives on an environmental issue and those of their teacher. Research on teachers' thinking in environmental education has made clear that teachers can hold strong views on the issues that they are teaching (e.g. Fien 1992; Kyburz-Graber 1999; Corney 2000; Cotton 2006a). We have seen in the chapter titled "Dealing with Emotions and Values" that the content of environmental courses can evoke strong feelings and emotional responses amongst students. It is therefore highly possible for students and teachers to hold strong and diverging views about the topics with which they are dealing. These situations of perceived differences of viewpoint are significant for two reasons. Firstly, they very rarely get surfaced by students within the learning situation but instead remain hidden and tacit. Secondly, as we will see from the examples below, they can have a very real effect on the nature and extent of students' engagement with environmental learning activities.

This kind of scenario was clearly seen amongst a group of engineering students in Lundholm's Swedish study. In the interviews about their ecology course, several students brought up the issue of the lecturer's perspective. Tobias, for example, talked about how he felt the course had been 'angled from an ecological perspective [i.e.] everything that humans do has an impact on nature and if you affect nature, it is bad'. He went on to explain how this conflicted with his own views on this topic:

> But, actually, humans are a part of the whole ecosystem too and therefore one has to live in harmony with nature. Humans did not used to do that, people in the Stone Age killed all the animals they saw. It wasn't all that environmentally friendly as one might think. [Compared to me, the lecturer] values environmental problems in one way and 'This is the right way'.

Tobias' views were echoed by one of his classmates, Sarah, who felt that it was 'a pity that the teacher was grousing at civil engineers, or it felt like it'. What she found challenging was the way in which the lecturer seemed to pass judgement on her and her classmates as trainee civil engineers: 'I think many of us got irritated because he kept on saying: You do this, and you and you. For example, 'This is something to think about when you build your roads'. But he doesn't know what kind of special courses I'm taking, or how I will work in the future'. Added to this, Sarah had the feeling that the lecturer's viewpoints were in fact hypocritical or at least inconsistent:

> Sometimes you kind of got the idea that he sort of lived in a little hood in an old fashioned way, but then you saw him driving his car instead of taking the bus! He had a car that uses a lot of fuel and emissions, and then it's a contradiction because he still wants that road. And, of course, there are problems, but you can't simply grouse and say bad things. He can't say that I'm destroying nature when he's doing the same.

In a similar way, another student called Ola reflected on the ecology course as follows:

> Our dear ecology lecturer has the viewpoint that man was God's biggest mistake. Humans have only destroyed and so on. Many of us [students] have got this impression and then it feels kind of meaningless to discuss how we can solve these problems if the best solution is if everybody kills themselves. This was crudely put, but it is a bit of this kind of atmosphere that has evolved. When somebody has asked a question about solutions to a certain problem you've got an answer that has been angled in that way. … He has these values that see man as an evil creature.

What is interesting in Ola's comments is what he has to say about the ways in which his engagement with the course is affected by his perception of the lecturer's perspective. For example:

> I think you quickly kill all interest in discussing, nobody wants to, what the hell I don't want to pick a fight. It is also this exam, I think that's a big part of this. If he has that opinion then you let him have it, then you don't go into discussion because you know that, even though it shouldn't be that way, the lecturer will be affected by discussions. I think, if you discuss the wrong things and have very different views, you don't want to become an enemy to someone who is going to correct your exam. There are a lot of values in a subject like this. I mean, just think about environmental issues!

What we see here are examples of how perceptions of the relationship between students' and teachers' environmental viewpoints can profoundly influence the learning experiences and responses of individual students. This shows how concerns about a lecturer's environmental views presented challenges for students, including one who decided not to express his real opinions on the topic for fear of 'becoming an enemy to someone who is going to correct your exam'. A similar, although far less marked, situation was seen with some of the students in Hopwood's study of English secondary school geography lessons (Box 1).

What all of the above examples show is the way in which perceptions of similarities or differences between students' and teachers' environmental views can

Box 1 Jenie's and Ryan's contrasting perspectives on a fair trade lesson

After a lesson about fair trade issues within a geography module on development, two students, Jenie and Ryan, made comments concerning their teacher's perspective.

For Jenie, the teacher's perspective on fair trade was influential upon how she experienced the lesson. As she explained in a post-lesson interview: 'I think Miss G is trying to get us to buy fair trade […] and maybe to make us think about things we do and things'.

By contrast, her classmate Ryan did not feel as if he had been affected in this way:

> A lot of people say at the end of the day it's going to change your opinion. It's not. It's not going to change your opinion, it's to help. It's like an instruction manual, you don't have to follow it but it's a guidance if you know what I mean.

influence the nature of students' experience of environmental learning tasks. While the issue of how teachers can and should deal with value-rich contentious topics within formal education has been debated for many years (e.g. School Council/ Nuffield Humanities Project 1970; Stradling et al. 1984), it is only recently that we have begun to understand some of the challenges that can arise for teachers and students grappling with environmental topics in everyday classrooms. Work by Cotton (2006b) has shown how teachers undertaking environmental education can experience significant difficulties in maintaining a balanced neutral role in the reality of the school classroom. In other words, while teachers may wish to conceal their views from their students, this is difficult to achieve in practice. Added to this, the important point emerging from our work (and other studies such as Schindler 1993 and Elliot 1995) is that environmental learning involves teachers and learners in negotiating viewpoints and the way in which this plays out in the minds of individual students can influence their willingness to engage in the activity at hand.

Differing Views of What Is Controversial

A second source of difference between students and teachers was where they held contrasting views on whether the subject matter is controversial. This is not about students and teachers holding different views on an environmental issue, as we saw earlier with the Swedish students and their ecology lecturer. Instead, it concerns students and teachers having different views of whether a particular environmental issue is controversial, topical and worth engaging with. There are parallels here with the Danish ENSI Project example discussed at the start of the chapter which involved student–teacher divergence over the type of animals worth focusing on for an extended piece of environmental project work. While the teachers had planned for a project focusing on *farm* animals, the students were expecting to work on *pet* animals (Elliot 1995).

A similar example from our research was the case of Laura, a 14 year old student, in a geography lesson about nuclear power (part of Rickinson's study). For Laura nuclear power is something she's 'not bothered about' and 'wouldn't sit around for an hour thinking full-on whether [it] is a good idea', while for her teacher it is something that is 'really affecting us, and we all need to sit up and take notice [and] consider the possibility that [a Chernobyl-type accident] could happen here'. This divergence of viewpoints between Laura and her teacher had a marked effect on the way in which Laura responds to the task of the lesson (Box 2). What is important, though, is that the divergence between Laura and her teacher was not to do with their views on the merits of nuclear power, but rather a difference of opinion over whether nuclear power constituted an issue that was worth thinking seriously about.

In talking further about this lesson, Laura explained how she had developed a strategy for dealing with this problem of having to express opinions about topics which she was not personally concerned about.

Box 2 Laura's difficulties with the nuclear power lesson

The students in this lesson had been asked to answer the question 'Do you think nuclear power is a good thing?' after watching a video about the Chernobyl disaster. Laura, a 14-year-old student, explained her difficulty with this task as follows:

> I could put 'No nuclear power is not a good idea', but I'd rather just write 'No' full stop. [...] It's not exactly hard to write what you think, but sometimes you're thinking, you don't write what you think, but you're thinking of something, but it's not what you think if you sat down and thought about it for ages. It's just a spur of the moment thinking for what would be right to put down. [I wouldn't want more time though...] because I don't find it interesting. I wouldn't sit around for an hour thinking full-on about whether nuclear power is a good idea, I wouldn't find it interesting, it would just be a waste of my time. Maybe not for miss [i.e. the teacher] or for other people, but that's generally what people think in our class.
>
> I just usually write what I think, just the truth...but sometimes I find it difficult, it depends. There's been some things on what you think and [looking back through her book] I remember what it was, about when there was that acid rain thing with the trees and miss said write what you think of what you see when you see those trees. I didn't really think anything so I just put that [pointing to what she wrote which said: 'When I see the effects of acid rain, I think it's just deserted with nothing much exciting, but I feel sorry for the trees'].…. I didn't really feel sorry for the trees, it was just something to write because that was what miss was trying to make us look at the effects the trees had from the acid rain. So I thought it was the thing to write down, it would make sense feeling sorry for the tree. Cos if these trees [pointing to trees outside window] were cut down, you wouldn't want them cut down so I just wrote that, but I wouldn't really care to be honest.
>
> [Researcher: Right, but you wouldn't write that down – that you don't really care?]
>
> No. Cos... miss wouldn't tell me off and she wouldn't say 'Well done', she'd just make me write it out again, she'd just have a conversation with me to see what I *really* think, so I just write that down and it saves me.

We see this as an example of a student engaging with a learning activity in a superficial or going through the motions-type way because of her belief that the topic of nuclear power is not worth thinking too hard about. In other words, a task that is designed by the teacher to encourage students to consider their own views on the issue of nuclear power is actually carried out by the student in terms of 'spur of the moment thinking for what would be right to put down'. For Laura, it was 'just the work we have to do' and not (as the teacher felt) something 'we've really got to grapple with'. There are echoes here of some of the issues reported by Schindler (1993) from the Austrian ENSI Project, in particular the fact that 'the pupils did not really consider the project to be "their" project' but rather saw it 'first and foremost as their teacher's plan [and] interpreted their task as to carry out orders' (p. 464).

The above situation, where the teacher's view of the subject matter as topical and controversial is not shared by the student, is completely reversed in the case of

Box 3 Lisa's environmental perspective on her population lessons

After studying demographic fertility and mortality rates, students in Lisa's class were asked to make a board game about population structure. No mention was made by the teacher during these lessons or in the task instructions of relationships between population numbers and environmental impact. However, Lisa described her board game as about 'how the environment is going to change through population', explaining that higher population numbers mean destruction of the countryside.

Similarly, Lisa discussed a series of lessons about migration as relevant to environmental issues because migration determines the number of people in a place and thus the extent to which the natural environment in that place is affected. In her words 'it's to do with population and how it affects the world'. Lisa added that it was this relevance to people–environment relations that made it worth studying reasons why people move from place to place.

another female secondary school student, Lisa, in a series of lessons about population trends (Hopwood's study). Here it is the student, rather than the teacher, who sees the learning activities as topically connected to environmental themes and so worth engaging with (Box 3).

In this example, learning experiences that were not intended by the teacher to be environmental in focus and controversial in nature were experienced and valued by Lisa in just these terms. The variability in students' individual responses is further emphasised by the fact that there were other students who were similarly uninspired by the topic of a lesson but, for other reasons, did in fact engage with the learning task. Jennie, a female secondary school student in Hopwood's study, for example, was very clear that she 'didn't give a damn about rainforests' in much the same way that Laura was 'not really bothered' about nuclear power. However, unlike Laura, Jenie participated enthusiastically in a lesson that involved a whole-class discussion on the topic of rainforests. Why? Because she liked debating: 'I like having my own opinion because I like arguing with people about other things, I like trying to put my view across. That's why I really liked the debating lessons because I get to argue'.

In summary, this section has shown how we need to be alert to potential differences not only in students' and teachers' views on environmental issues (as in the preceding section), but also in students' and teachers' views on whether topics are controversial. The examples of Laura and Lisa have shown how student–teacher differences on this can go both ways. On the one hand, students can engage only superficially where, in contrast to their teacher, they fail to see real controversy in the topic they are studying. On the other hand, it can be the student, rather than the teacher, who sees a particular learning activity as topically connected to environmental themes and so worth engaging with. In addition, however, there can be students who are uninspired by the topic of a lesson but, for other reasons such as a love of debating, do in fact engage with the task at hand. Taken together, all of these examples underline the complexity of relations between teachers' and students' views about environmental curricula and the importance of understandings of environmental learning that are sensitive to such dynamics between learners and teachers.

Differing Views of What Is Relevant

The intricacy and variety of students' views on the relevance (or otherwise) of their environmental learning has been very clearly illustrated in the chapter titled "Questioning Relevance". We have seen how learners can have strong beliefs about what counts as relevant both to them now and in the future and to particular subjects or courses. Given that teachers can also hold strong views on what is and is not relevant to study in a particular subject, it is not surprising that diverging views on relevance was another example of student–teacher differences within our studies.

A secondary school geography lesson about the Kayapo Indians of the Amazonian rainforest provides a helpful example from Rickinson's study. For the teacher, it was very clear that learning about the indigenous peoples of the rainforest was appropriate and useful for a Year 8 geography lesson. He felt strongly that students needed to understand both the cultural and the physical characteristics of the rainforest and this meant studying its peoples as much as its animals and thinking about development as much as environment. The Kayapo lesson, therefore, came within a 12-lesson module on Rainforests that began by looking at physical features and climatic characteristics before exploring rainforest peoples and cultures and eventually the development and destruction of rainforest environments.

When the experiences of individual students in this lesson were probed, however, it became clear that the teacher's views on the importance of learning about Amazonian people as well as Amazonian vegetation/climate were not necessarily shared. One boy in the class, Aiden (13 years old), felt that learning about the peoples of the rainforest was 'not really geography'. When interviewed shortly after the Kayapo lesson, he said: 'In geography today I did not learn anything to my benefit. Yesterday's lesson on climate was much more interesting because you were actually learning something about the rainforest, not the people who live there'. When asked to elaborate on this, it became clear that this was related to his view of what was legitimate content for a geography lesson:

> Today we were mainly focusing on the people, and not the rainforest, and their habits in the rainforest. But that's not really learning about the rainforest, it's about learning things that people do in the rainforest. It's like you're learning architecture and learning what people can do in the house in many ways. You're learning how to build a house, and they're telling you what people can do in this house. It's not really talking about the architecture of the house.

Far from being geography, Aiden felt that 'You probably would have learnt about the Kayapo in history as history is about things that happened and people in many ways'. This was not at all the case for Joanne, another student in the same class, who had no particular difficulty with the relevance of the subject matter and quite enjoyed the lesson because of what it involved in terms of activities: 'It was a nice change to watch the video and I enjoyed doing the poem'.

Aiden's negative response to the Kayapo lesson is not unlike some of the students described in Schindler's (1993) Austrian case study. In this case it was the nature of the task rather than the content that was problematic for the students. The Austrian 13 year olds found the experience of independent project work very different to the 'activities they were used to in normal teaching' and were unsure whether it was 'really work' and consequently made little progress (p. 466). Here again, then, we see evidence of important differences between teachers and students ideas of what is appropriate to study and do in an environmentally-focused lesson.

In Lundholm's university-based study there was also evidence of students experiencing difficulties with the nature and structure of their environmental courses. An issue for the engineering students, for example, was whether the course content was relevant to them in terms of working as an engineer in the future. Their complaints were that the content of the course, as put together by their ecology lecturer, was (i) too focused on problems as opposed to solutions ('This ecology course ought to give solutions to problems and not only the problems'), and (ii) overly concerned with the big picture ('Let's say the 'water cycle. ... you don't really need the big picture, it's enough to know that in this place water runs from here and on').

The common theme here is of students expressing personal evaluations of the relevance of environmental content and the way it is organised that conflict with the starting points of their teachers. This shows that one cannot assume that students will share teachers' beliefs in the importance of dealing with environmental issues within particular school subjects or university disciplines. Rather there is a need for teachers and curriculum developers to think carefully about the ideas that students bring to the learning situation about what is and is not relevant and helpful to study in different curriculum contexts.

Differing Views of Empathy Tasks

Studies of school teachers involved in environmental education have found that presenting students with 'a balanced picture of controversial environmental issues' is a common aspiration (Cotton 2006a, p. 77; see also Corney 2000; Gayford 2004). In other words, it is frequent for teachers to emphasise the importance of taking a balanced or neutral role, providing access to a range of views and not stating one's own views. Empathy tasks, such as role plays or empathetic writing activities, are a type of learning activity that is often used in connection with these kinds of goals. As Stradling et al. (1984, p. 113) reported some time ago concerning the teaching of controversial issues in school, 'empathetic procedures are often turned to when the issue involves a group which is unpopular with some or all of the students, or when the issue involves covert discrimination against a particular group or the issue is remote from students' own lives'.

When these kinds of tasks are looked at from the perspective of teacher–student negotiation, it becomes clear that they can present difficulties for learners. Empathetic role play, for example, can be challenging if you are very keen on expressing your own perspective in debates but are asked to play the part of a perspective that is quite different from your own. By contrast, an empathetic writing exercise like writing a poem from the perspective of someone very different to yourself can easily turn into an exercise in creative, but not empathetic, writing if you are a student who really likes writing poems that rhyme. More detail about these examples is given below, but the important general point is that empathy tasks can present very particular kinds of challenges for students' environmental learning and can result in learning processes that differ considerably from teachers' original intentions.

One example of this can be seen in Box 4 which describes events within a school geography lesson in which the students were carrying out a group role play activity about rainforest development (Rickinson's study).

On the face of it such an interaction could well be seen as a fairly regular part of classroom life where students and teachers are continually challenging each other in any number of ways. When the teacher and student were interviewed after the lesson, however, it became clear that this interaction was more than a passing spat. Instead it represented a genuine difference of perspective between the student and teacher. Simon reflected on the lesson as follows:

Box 4 Simon's difficulties with the rainforest role play

Picture the scene: the students, sitting in groups, have just received their role cards representing differing perspectives on rainforest development. The teacher is moving around the room from group to group when the following interaction with a male student, Simon, occurs:

Simon: I don't want to be someone against the development, I want to be someone for it
Teacher: Well see how you do with that, it's important to be able to look at another viewpoint Simon
S: I'm against stopping it, so how come?
T: I know you are, but it's a chance to see another viewpoint
S: I can see it, but I can't agree with it
T: I'm not asking you to agree with it, I'm asking you to put across that view
S: I'm not saying anything then
T: I think you're going to find you're going to have to
[Teacher moves away to deal with other student groups]

> Last lesson I thought was really boring. I just don't like, I like keeping my view and saying my view rather than using someone else's. I mean I can do it and I have done it but I just don't like doing it. I just like to stick to my view [...] I'd say that I didn't really enjoy it [...] because basically I wasn't speaking about my view. [...] [If I could have been representing my own view] I think I'd have enjoyed it more because it would have been my view, and I would have actually had what I thought as well as what was written down.

By contrast, Simon's teacher felt very strongly that, where possible, role play should involve students taking on roles with which they do not necessarily agree with. In her words, 'I just think it's good for them to have to get into a different viewpoint'. She expanded on the value of empathetic role play as follows:

> It's part of developing their understanding for them to see different perspectives so that they can't just dismiss the Indians, the charcoal burners or the rubber tappers or whoever without trying to understand why they do it and what it must be like to be there. Role play can do this because they're having to speak with the voice of that person, and they're much more likely to understand where that person is coming from if they're having to be that person.

This example provides a glimpse of the complexity that can be involved in undertaking role play exercises. This is just one student's response to one part of one lesson but it serves to highlight the way in which these kinds of tasks can present particular kinds of challenges for student engagement with environmental content.

Another kind of learning activity where we have seen interesting teacher–student divergences is with empathetic writing exercises. Take the example of a secondary school geography lesson about indigenous rainforest peoples that was part of Rickinson's study. The teacher asked the students to 'try and put yourselves in the position of a Kayapo Indian who doesn't have a clue about what it's like in England, who only knows where they live in the Amazonian rainforest'. The task, after watching a short documentary film about the Kayapo Indians, was to write a poem about 'a day in the life of a Kayapo'. Box 5 shows the poems that two students, Joanne and Aiden, wrote in this lesson.

Through talking with these students after the lesson, it became clear that this empathetic writing task had been understood quite differently by Joanne and Aiden. For Joanne, it was an exercise in creative, as opposed to empathetic, writing, where her main concerns were the rhyming of words and the rhythm of her poem. This was how she described the process of writing her poem:

> It was a bit harder because I thought it would be easy and I started off really well and then I started, like no words left, no like words and I started making up a few [...and...] when I got things like [reading from her poem] 'I live in the grass along by the stream, My face is painted yellow and green', and then I couldn't think of anything else 'cos that line, when you put it with that it didn't go because you'd be like this [acts out a gasp] gasping for air.

Aiden, meanwhile, was more concerned with putting 'some sort of content in it' and trying to include information about what 'they might think'.

> Well I wanted some sort of content in it, so I put that in....I just put that in there to sort of put it in as a view, you know they might think. Well, they're not using CS gasses or anything like that, are they? They're just using what they have in the forest but whereas we just, well we buy it in gases, we use chemicals and so on [...] I just thought it would be good since I was writing a poem from their view and what they would probably think.

Box 5 Joanne and Aiden's poems about the Kayapo Indians

Joanne	Aiden
I'm a Kayapo you know what I mean I live by the grass along the stream In my village, where I live with my dad and my uncle Viv I go fishing with my dad all my mates think I'm sad. The tribe was started long ago by a couple of people all named Joe	In the undergrowth of the forest floor Wild boar roam free again As our tribe is environmentally- friendly the Western people corrupt and pollute. As we fish for food the women prepare vegetable stew. While the children practice their hunting skills, the men do the real thing. The Western people want to change us But we will not change our ways. We are the Kayapo strong and true.

Compared with Joanne, Aiden's interpretation of the task was much closer to the original intentions of the teacher. What is interesting here is the way in which the task acted as a barrier rather than a facilitator to Joanne's engagement with the subject matter. In other words, she was caught up with writing a poem that rhymed, rather than thinking about the lives of the Kayapo Indians. There are interesting parallels here with studies of students' responses to writing tasks in other curriculum contexts. Doyle (1986 p. 376), for example, described the difference between two students' understanding of a written assignment in a junior high school English/social studies class in terms of one adopting a 'procedural interpretation' and the other a more 'substantive interpretation'. The task was a written assignment about the American Indian tribes of the region. The difference between the two students in Doyle's study was that one 'saw the work as essentially a process of turning notes into paragraphs [and] had little interest in, or knowledge about, Indians and did not appear to think that gaining such knowledge would be helpful' (procedural, and similar to Joanne), while the other was much more concerned with gaining information 'about a real Indian tribe that she could use in writing her paragraphs' (substantive, and similar to Aiden). Connections can also be made with what Edwards and Mercer (1987) termed 'ritual understanding' and 'principled understanding'. They proposed that students' understanding can be 'either principled (understanding the issues and concepts and their relationships to the activities) [as with Aiden] or ritual (embedded in the paraphernalia of the activities themselves, without any grasp of what it was about) [as with Joanne]' (p. 99).

This, then, is another example of the way in which empathy tasks can present challenges for some students' environmental learning. Overall in this section, we have seen another way in which student–teacher differences can play out in environmental learning situations, that is, in the context of tasks designed to develop empathy. Simon's difficulties with playing the part of 'someone against the development' highlighted one way in which empathetic role play can be challenging for learners. Meanwhile Joanne's poem about the Kayapo showed how the task of writing an empathetic poem can turn into an exercise about rhythm and rhyme rather than indigenous cultures and lives. Both of these examples underline the complexity of the demands that are placed on learners and teachers involved in empathy-based tasks. As Schindler (1993) observed for independent project work, it is all too easy to disregard the skills students need to cope successfully with these kinds of tasks.

Summary

This chapter has shown how a lens focused on the viewpoints of students and teachers can reveal new dimensions and layers of complexity within environmental learning situations. We have seen various ways in which classroom-based environmental learning is characterised and shaped by an intricate interplay between the viewpoints of students and teachers. This interplay sometimes manifests itself within the explicit action of the classroom such as a student actively objecting to a particular task but more often seems to remain hidden and implicit within students' individual thoughts and task interpretations.

Overall we have seen four ways in which environmental learning can involve tensions between students and teachers. These concern learners and teachers holding differing viewpoints on: environmental issues, what is controversial, what is relevant, and empathy tasks. Using the framework outlined in the chapter titled "What Is Environmental Learning?", we can see how these student–teacher differences relate not only to the *what*, but also to the *where* and the *how*, of environmental learning.

In relation to *what* is learnt, it is clear from the Swedish engineering undergraduates that perceived differences between students' and teachers' environmental viewpoints can profoundly influence the experiences and responses of individual learners. Think, for example, of Ola deciding not to express his real opinions during his ecology course for fear of 'becoming an enemy to someone who is going to correct your exam. Another potential difference relating to the content is where students and teachers hold diverging views on whether the subject matter is controversial. Laura's view of Chernobyl/nuclear power as 'just the work we have to do' as opposed to (in the words of her teacher) an issue 'we've really got to grapple with' was a key example here.

Student–teacher tensions and differences were also evident for the *where* of environmental learning. There were clear examples of teachers' beliefs about the appropriateness of dealing with particular kinds of environmental content in particular curriculum contexts not being shared by individual students. Some of the Swedish engineering undergraduates were far from convinced that their ecology course was helpful to them in terms of working as an engineer in the future, while the English school student Aiden felt strongly that learning about the indigenous peoples of the rainforest was 'not really geography'.

Finally, in relation to the *how* of environmental learning, this chapter has highlighted complexities and challenges in environmental learning activities focused on empathy. Empathetic role play, for example, can be challenging if like Simon you are very keen on expressing your own perspective in debates but are asked to play the part of a perspective that is quite different from your own. By contrast, an empathetic writing exercise like writing a poem from the perspective of someone very different to yourself can easily turn into an exercise in creative, but not empathetic, writing if like Joanne you are a student who really likes writing poems that rhyme.

The significance of all of these examples of student–teacher differences lies in the very real effects they can have on the nature and extent of students' engagement with environmental learning activities.

Enhancing Environmental Learning

In this final chapter, we re-visit the main arguments and findings of the book and consider what these tell us about students' environmental learning in formal settings. We then discuss how the ideas presented in this book might inform the development of environmental learning in terms of practice, policy and research.

Overview of Main Arguments and Findings

The key motivation for writing this book is the urgent need for more and better research-based understandings of environmental learning and students' experiences. In our early chapters, we show how the learning demands associated with environmental and sustainable development issues are becoming ever greater, but research into the nature and dynamics of such learning is still in its early infancy. Despite the fact that environmental education and education for sustainable development have become features of many countries' formal education systems, very little is known about what such provision looks and feels like for the learners concerned. Until recently, there have been all too few empirical investigations specifically focused on learning in environmental education. The tendency has been to focus on environmental learning outcomes as opposed to environmental learning processes and to make little use of wider learning theory.

This situation is changing, however, as more authors have sought to place learning at the centre of discussions about environment and sustainability. Books such as *Sustainable Development and Learning* (Scott and Gough 2003), *Social Learning Towards a Sustainable World* (Wals 2007), *Participation and Learning*: *Perspectives on Education and the Environment, Health and Sustainability* (Reid et al. 2008), and *Engaging Environmental Education*: *Learning, Culture and Agency* (Stevenson and Dillon forthcoming) all reflect an increasing emphasis on learning. Coupled with this has been the emergence of a small number of empirical research studies focused on the subjective experiences of learners as they are engaged in various forms of environmental education.

M. Rickinson et al., *Environmental Learning: Insights from Research into the Student Experience*, DOI 10.1007/978-90-481-2956-0_8,

This book is part of these developments. Through in-depth examination of school and university students' experiences and interpretations of environmental lessons or courses, the findings of this book add new detail to current knowledge about environmental learning in formal settings. The specific examples and wider discussions in the chapters titled "Dealing with Emotions and Values", "Questioning Relevance" and "Negotiating Viewpoints Among Students and Teachers" provide detailed illustrations of how environmental learning plays out amongst students and teachers in everyday classrooms. These help to highlight a number of characteristics of environmental learning (see below) which, while not uncommon in the wider literature on learning processes, have not been well evidenced in the context of environmental education.

The Active Role of the Learner in Environmental Learning

Perhaps the clearest message emerging from the preceding three chapters is the sense of learners as active agents in environmental learning situations. Thinking back to our earlier-discussed assumptions about the nature of learning (as involving active participation rather than passive receipt of knowledge), we find indeed that learners play a significant role in shaping the process of environmental learning. The preceding three chapters have demonstrated a wide range of ways in which learners mediate the nature, focus and shape of their environmental learning. The students in our studies were powerful filterers of environmental content and tasks in terms of what they attended to, what they saw as relevant, what they ruled in and ruled out, what they did, how they did it and so on. All of this underlines the importance of what students bring to the learning situation in terms of ideas, preferences, interests, value positions, emotional concerns and viewpoints. These influences all play out within the learning process through a range of in-the-moment judgements students make about relationships between themselves, their peers, their teachers, subject matter, tasks and learning outcomes.

The Centrality of Values and Emotions in Environmental Learning

Another recurring theme in the preceding three chapters is the critical role that values and emotions play in the dynamics of environmental learning. Our work has highlighted the importance of the affective in: (i) students' responses to environmental learning activities ('I don't like animals being hurt or moved so I was hardly watching the video'); (ii) students' perceptions of environmental subject matter ('Geography is more a kind of opinion subject, you can't be right or wrong because it's your opinion'); (iii) students' encounters with new concepts and disciplines ('In economics everything has to be shown in dollars and cents when a decision is made and my worldview really opposes that'); and (iv) student–teacher relations and interactions ('I think Miss G is trying to get us to buy fair trade and maybe to make us think about things we do'). These examples show how environmental subject matter and tasks can provoke strong emotional reactions and challenge learners' closely-held values. Students are therefore grappling with their own affective responses and their ideas of what role these could or should play in a formal learning environment such as a school classroom or university lecture course.

The Potential for Student–Teacher Tensions in Environmental Learning

The research reported in this book has also flagged up the complex interplay that can take place between the viewpoints of students and teachers during environmental learning. This interplay sometimes becomes manifest within the explicit action of the classroom such as a student actively objecting to a particular task or idea, but more often seems to remain hidden and implicit within students' individual thoughts and task interpretations. Whether implicit or explicit, student–teacher differences can significantly affect the course of environmental learning in classrooms. We saw examples of this in our work where students and teachers had differing views on: environmental issues ('Our dear ecology lecturer has the viewpoint that man was God's biggest mistake'), what is controversial ('I wouldn't sit around for an hour thinking full-on whether nuclear power is a good idea'), what is relevant ('Learning about the people of the rainforest is not really geography') and empathy tasks ('I like keeping my view and saying my view rather than using someone else's'). Tensions like these can play out in different ways for different learners, but can lead to disengagement, frustration, or sometimes confusion as to what is expected or deemed appropriate by the teacher.

The Complexity of Students' Experiences in Environmental Learning

It is also clear from the findings in this book that the perspectives and experiences of learners involved in environmental education are highly individualised and nuanced. One source of complexity is variability between learners. Each of the preceding three chapters illustrated a wide variety of learner experiences and responses even within very specific settings such as a particular school class or a small group of university students working on a joint project. On the theme of relevance there were intricate differences between the positions of individual learners, such as the three female secondary school students who saw environmental learning as most relevant to them when it was about 'places where I live' or 'environments of which I have little knowledge' or 'things I feel strongly about'. As well as variability between learners, we have also seen complexity within the views held by individuals. For example, learners can feel removed and distant from the issues at hand, but at the same time feel that it is personally important for them to learn about them. Similarly, they can find aspects of content unappealing, but recognise their potential importance in their imagined future professional roles. In summary, the process of environmental learning appears to be a highly personal one.

The Multi-layered Nature of Environmental Learning and Teaching

Finally, the general impression that emerges from this work is of the multi-layered nature of environmental learning and teaching. When one considers the richness of students' accounts and the variety of their experiences and responses, it is clear that there is a lot going on in environmental education classrooms. Small differences in the nitty gritty of environmental learning situations, such as two students taking contrasting approaches to a task, can in fact be underpinned by significant differences of perspective and response among learners. Indeed, the reason that insights into students' experiences matter is because time and again we have seen how they have a real and tangible effect on the nature and quality of students' environmental learning.

In conclusion, it is important to stress that the above findings represent the early beginnings of research-based understandings of environmental learning in formal settings. There is clearly much more work to be done in this area (see implications for research below). That said, we are confident that the findings in this book are sufficiently robust and well grounded in the actualities of classrooms across a range of settings to be able to raise useful questions for the development of future practice and policy. The phrase 'raise useful questions' is used deliberately to signpost our view of any implications as issues for consideration as opposed to strategies to implement. In other words, they are based on a conceptual (research raising questions), as opposed to an instrumental (research providing answers), understanding of research use (Estabrooks 2001; Rickinson and Reid 2003). As Nutley et al.'s (2007) analysis of research use across a range of sectors (education, social care, health care, criminal justice) has shown:

> [While] much attention has been focused on instrumental use of research – where research evidence has concrete and visible impact on the action and choices of policy-makers and practitioners [...] We know that, on the ground, research is often used in more subtle, indirect and conceptual ways [...] altering the ways in which policy-makers and practitioners think about what they do, how they do it and why (p. 301).

Enhancing Environmental Learning Practices and Policies

We see important messages stemming from this work for teachers, lecturers and other educators seeking to facilitate environmental learning amongst students. Perhaps the most important implication is the need for environmental education practitioners to be sensitive to the potential challenges and complexities of environmental learning situations. This is about recognising the ways in which: (i) environmental learning can be difficult for learners; (ii) environmental learning experiences can vary between learners; and (iii) environmental learning can involve tensions between students and teachers.

Our research has highlighted a number of difficulties and complexities that can be encountered by students as part of school and university-based environmental learning. While more work is needed to clarify the prevalence of these in other formal and non-formal learning contexts, we still see them as raising potentially powerful questions for practitioners currently involved with environmental education in its many guises. As shown in Table 1, questions for practitioners can be identified from many of the specific difficulties or complexities discussed in this book. Taken together, the emerging questions are an invitation for environmental educators to explore whether any of the issues discussed in this book could apply in some way to the learners and learning contexts they are working with and, if so, what might be done in response.

We recognise that what is implied here is not a straightforward undertaking but one that requires genuine engagement and reflection on the part of practitioners. As shown by studies of teachers' responses to students' ideas about teaching and learning, the process involves teachers in both 'comfortable and uncomfortable learnings' (McIntyre et al. 2005). That said, we firmly believe that the kinds of questions raised in Table 1 need to be part of practitioners' efforts to enhance students' environmental learning. To this end, professional learning opportunities for environmental education practitioners are important in two key areas: understanding the emotional dimensions of environmental learning and accessing students' learning experiences.

Understanding and Negotiating the Emotional Dimensions of Environmental Learning

The need for teachers and other educators to take account of the cognitive/conceptual challenges inherent to environmental issues is well recognised (e.g. Boyes and Stanisstreet 1996). But our work suggests that it is equally important for environmental education practitioners to be able to foresee the potential affective/emotional challenges that may arise when dealing with particular topics/learning activities. This calls for skills in what some teacher education researchers have termed 'emotional scaffolding', which is about how teachers can 'help students to build emotional, as well as cognitive, relations to what they are learning'

Table 1 Students' environmental learning difficulties or complexities and their possible implications for environmental education practitioners

Difficulties or complexities for learners	Examples from our research	Emerging questions for practitioners
Some aspect of the topic triggers discomfort for students which causes them to disengage on some level	'I don't like animals being hurt or moved so I was hardly watching the video'	• How could such responses be anticipated and recognised? • How could I draw on such responses as part of the learning?
Students have emotional difficulties with certain key concepts in the subject	'In economics everything has to be shown in dollars and cents when a decision is made. My worldview really opposes that'	• What emotional difficulties might my students have with subject matter? • How could I support students who experience such tensions?
Students feel that certain topics are more valuable than others and this affects their engagement and enjoyment	'In geography today I did not learn anything to my benefit. [...] We were mainly focusing on the people [in the rainforest] but that's not really learning about the rainforest'	• How much do I know about my students' different ideas of relevance? • How do they compare with mine and what could I do differently in response?
Students are unwilling to engage in class discussions due to concerns about opposing their teacher's viewpoints	'You quickly kill all interest in discussing ... you don't want to become an enemy to someone who is going to correct your exam'	• How could my students be affected by perceptions of my values? • How could this be handled more productively?
Unlike their teacher, students don't see the topic as in any way controversial and struggle with taking a stance or expressing a view	'[Nuclear power is] just the work we have to do [...] I wouldn't sit around for an hour thinking full-on about whether nuclear power is a good idea, I wouldn't find it interesting'	• To what extent are my students 'going through the motions' in their learning? • How can I help students to do more than this?
Unlike their teacher, students object to role playing views that are very different to their own	'Last lesson I thought was really boring. I just don't like, I like keeping my view and saying my view rather than using someone else's.'	• What difficulties do my students experience with empathy tasks? • How could I help them to negotiate these?

(Rosiek 2003, p. 410). A potentially helpful distinction connected with this concept is that of 'constructive emotions' (that serve to focus student attention more closely on the salient aspects of the subject matter) and 'unconstructive emotions' (that serve to distract students from the subject matter content or in some other way inhibit their learning). We have seen several examples of the latter in the environmental learning contexts discussed in this book. In view of this, we would suggest

Table 2 Strategies for negotiating students' emotional responses to the subject matter

Approach to emotional scaffolding	Attempts to foster constructive emotions about the subject matter	Attempts to reduce unconstructive emotions about the subject matter
Implicit	An effort is made to foster a constructive emotional response to the subject matter by associating it with something students find familiar or interesting	An effort is made to avoid triggering an unconstructive emotional response to the subject matter by approaching it in an unfamiliar context
Explicit	An effort is made to foster a constructive emotional response to the subject matter by drawing attention to it and offering students reasons why the effort to learn it is worthwhile	An effort is made to avoid triggering an unconstructive emotional response to the subject matter by drawing attention to these emotions and making light of it or by assuring students it is 'not as bad as it seems'

Source: Rosiek (2003, p. 407).

that environmental education practice could benefit from consideration of the strategies outlined in Table 2 (Rosiek 2003) and what these might entail for different forms and contexts of environmental learning.

Accessing and Understanding Students' Learning Experiences

Our work has also highlighted the often hidden nature of the student learning experience in formal environmental education. In the case of student–teacher differences, for example, students may have good reasons for wishing to keep their differences of opinion or views about the subject matter hidden from their teacher. The second important area for practitioner professional development, therefore, concerns accessing the experienced curriculum. This is about practitioners developing: skills in consulting students about aspects of their environmental learning; belief in the contribution that learners' ideas can make to the enhancement of environmental education practice; and open-ness to changes in the balance of power between learners and educators. Use of lenses similar to the ones discussed in this book (Fig. 1 of the chapter titled "Lenses for Understanding Environmental Learning") may well be helpful in framing and focusing such consultation processes. Likewise, the challenges and complexities experienced by the students in our studies (Table 1) may provide some useful starting points for discussion and reflection. The important underlying argument, though, is that improvements in students' environmental learning will be far more likely where educators recognise the importance of, and are routinely involved in, accessing and understanding their students' learning experiences.

The above points are not simple tips for practitioners. They are about developing new understandings of environmental learning, new skills in environmental teaching and new forms of engagement between learners and teachers. These are not tasks for environmental education practitioners in isolation, but rather need to be seen in connection with wider policy frameworks relating to environmental education and education for sustainable development. In particular, we see implications for policy-makers with responsibilities for environmental and sustainability education in formal setting such as schools, universities and teacher education institutions. There are also messages for education policy-makers in non-governmental organisations that are developing resources and training opportunities for practitioners in this area.

Most importantly, the findings of this book highlight the shortcomings of environmental and educational policies that are based on simplistic, straightforward and unproblematic views of school-based environmental education. Policy development in relation to sustainability needs to take into account the complexity of teaching–learning processes and the important differences between the espoused, the enacted and the experienced curriculum. The preceding three chapters have highlighted the kinds of challenges that learners and teachers can face in seeking 'to weave [sustainability] issues into the curriculum' (UNESCO 2005, p. 19). Goals such as making learning 'locally relevant', 'interdisciplinary' and 'participatory' and developing learning experiences 'based on the principles and values that underlie sustainable development' are no small tasks for teachers and students in everyday learning contexts (UNESCO 2005, p. 30). Our work, along with many other studies, reinforces the critical need for policy developments in this field to be supported by strategic investment in initial teacher education, continuing professional development and practitioner research and development projects that take seriously the pedagogical challenges involved with environmental and sustainability education.

Enhancing Environmental Learning Research

Closely connected with these issues for practitioners and policy-makers are a further set of implications for those involved in research. It is important to remember that learners have tended not to feature strongly in environmental education research agendas. Environmental learning remains relatively little researched, weakly evidenced and poorly theorised. With this in mind, we see five priorities for further research.

Further Research on Environmental Learning and Learners' Experiences

If nothing else, we hope that the examples and findings contained with this book serve to heighten readers' curiosity about learners and learning in environmental and sustainability education. There is an urgent need for questions about learning processes and learning experiences to become a more routine part of programme evaluations and research inquiries conducted in the field. As Reid and Scott (2006a, p. 243) have argued, 'merely focusing on education (as a process with pre-figured outcomes) without a complementary focus on the learners themselves and what they want to learn is perverse'. This is not about focusing on process and experience issues in isolation or seeing them as much more important than other potential foci. Rather it is about ensuring that environmental education evaluation and research work towards a better balance between: (i) the perspectives of learners and the perspectives of educators; (ii) questions about learning processes and questions about learning outcomes; and (iii) concern with the experienced curriculum and concern with the specified/enacted curriculum. Alongside greater attention for these kinds of issues within the field of environmental education research, it is also important that we work towards environmental learning becoming better represented in research programmes focused on learning more generally. We are conscious that environmental learning has been virtually absent from the projects and portfolios of recent major educational research initiatives such as the UK's Teaching and Learning Research Programme (and similar programmes in Finland, Norway and New Zealand).

Greater Emphasis on Emotions and Values in Environmental Learning

The examples and findings discussed in this book have demonstrated various ways in which emotions and values can feature strongly in environmental learning situations. What we have seen are complex interactions between the value-laden nature of environmental subject matter, the affective dimensions and challenges of environmental learning activities, the emotions and values of learners and the emotions and values of teachers. The findings presented in this book, though, are an early step towards understanding the dynamics of such interactions. We need more detailed investigations of the affective dimensions and emotional dynamics of everyday environmental learning and teaching situations within formal, informal and non-formal contexts. There is potential here for productive connections with developments in research and theory across various areas of education. It is clear that researchers in many fields are now recognising 'the role of feelings and emotions in the learning process [...as...] a new and largely unexplored area' (Efklides and Volet 2005, pp. 377–379).

Better Use and Development of Theory in Environmental Learning

Moving towards more sophisticated understandings of issues such as the emotional dynamics of environmental learning will require not just more empirical attention but also better use of wider knowledge and theory about learning. As we and other have argued, engagement with learning theory and the development of conceptual models have not been strengths of environmental education research (Dillon 2003; Myers 2006). This is problematic because 'narrowly focused and limited views of environmental learning limit both its understanding and a meaningful assessment of its impact' (Falk 2005, p. 273). As argued by Hart (2007, p. 31), 'learning is not what it used to be' and new perspectives on what counts as learning can raise important questions and ideas for understanding environmental learning and participation (see also Reid and Nikel 2008). We hope that our use of lenses focused on particular dimensions of environmental learning (see the chapter titled "Lenses for Understanding Environmental Learning" and Appendix II: Development of the Lenses) has demonstrated the value of combining in-depth empirical investigation with analysis of wider research and theory.

Broader Consideration of Life-Long Environmental Learning Contexts

What we have presented in this book are early sketches of environmental learning processes and experiences within particular kinds of formal education settings (secondary school environmental geography classrooms and university environmental courses for engineers and biologists in England and Sweden, respectively). A key challenge is thus to procure empirically-robust insights into environmental learning in a wider range of contexts than those focused on here. We know little about what lies beneath the surface of environmental learning in other formal settings (different countries, age ranges, subject or curricular) and, beyond that, within informal and non-formal learning contexts. The challenge of sustainability is one that necessitates learning at all levels within society: 'the learning that will need to be done transcends schools, colleges and universities; it will be learning in, by and between institutions, organisations and communities' (Scott and Gough 2003, p. xiv). In view of this, as we have argued elsewhere, environmental learning research needs to be both life-wide and life-long (Rickinson 2006). This means using a life-course perspective to think about what we know and what we need to know about environmental learning during infancy, childhood, adolescence, adulthood, middle age, retirement and old age (e.g. Pollard 2003).

Stronger Collaborations Between Researchers, Practitioners and Learners

This book has highlighted the often hidden challenges that students can experience during environmental learning activities and the importance of educators being able to access and understand their students' learning experiences. Progress in this area, though, will require much closer collaboration between researchers, educators and learners involved in environmental learning. Building on processes started in the

OECD Environment Schools Initiative (ENSI) Project (Elliot 1995; OECD 1995), we now need more research and development projects that look not only at students' perspectives on environmental learning but also how these ideas can be used by educators and learners to improve future environmental learning activities (see McIntyre et al. 2005 for a school-based example of this kind of work). To this end, there could be real value in considering design-based methodologies such as 'design experiments' (Cobb et al. 2003) and 'development and research (D&R) projects' (Stanton 2006) as possible approaches for combining practical improvements with theoretical development. The underlying point here is that the task of researching and understanding environmental learning is not solely one for those who see themselves as researchers.

This final priority area concerning stronger collaboration underlines the fact that enhancing environmental learning research is not just about research topics but also very much about research approaches. The shaping of future environmental learning research needs to take careful note of debates and developments in education and social science research more generally. Against the backdrop of evidence-based policy and practice discourses, there are increasingly international efforts to improve the relevance, quality, coherence and usefulness of research in education and other social sciences (European Commission 2007; OECD 2007). Such developments have become 'closer and more frequent realities' for research in environmental education and can be seen to raise important and difficult questions about the scale, reach and methodological tendencies of current work in this area (Reid and Scott 2006b, p. 571). Developing enhanced research and research-informed practices in environmental learning therefore needs to be part of wider efforts to improve the generation, communication and utilisation of research evidence in relation to education and sustainability.

Appendix I: Empirical Context

This appendix provides contextual information about the three studies which are the basis for this book. In each case we offer methodological details regarding sampling and data collection techniques, and describe the contexts in which fieldwork was conducted (information about school/university sites, outlines of lessons or courses observed and discussed with participating students).

Rickinson Study

Methodological Details

This study sought to address the following research questions:

- How are geography teachers teaching environmental curriculum topics to their Key Stage 3 classes (ages 11–14 years)? Why are they teaching in these ways?
- How are their students experiencing the subject matter-task aspects of these lessons?
- How are the teachers' and students' views of the subject-matter task aspects of such lessons similar and/or different?

The key methodological task was one of generating authentic accounts of teachers' and students' thinking about their teaching and learning during environmental geography lessons. To this end, the adopted methodology was qualitative, being small in scale (focusing on the teacher and four students from one Year 8/9 class in three successive schools) and grounded in approach (following an emergent methodology and a progressive focusing of substantive concerns). Features of particular importance included: a focus on classroom practice and students' and teachers' experiences of this; a concern to generate authentic accounts from both students and teachers; and recognition of the importance of researcher-research participant relationships.

The research comprised three sequential case studies (each in a separate school) of the teaching and learning of environmental subject matter within a small number of Key Stage 3 geography lessons. Each of these case studies involved spending around 2 months in the school attending all the geography lessons of the selected class/teacher (usually 2 per week) during the teaching or all, or part of, an environmental module.

The selected teachers represented a purposive sample of three geography teachers who were:

- Teaching what they perceived to be 'an environmental geography module' to a Key Stage 3 class
- Interested to be involved in such a research project
- Working in schools which were easily accessible.

The first criterion was clearly driven by the focus of the study, the teaching and learning of environmental geography subject matter at Key Stage 3. The second criterion reflected my belief that teachers with a genuine interest in my research would be the ones most likely to yield the rich insider accounts about practice that I was seeking. This was related to the desire to work with a small number of teachers. It also recognised the considerable time commitment that needed to be made by the participating teachers, which, it was felt, would be more realistic with genuinely interested individuals. The third criterion was related to the practicalities of carrying out a study which necessitated frequent school visits at varying times over a prolonged period.

The selection of the class that would be worked with, and the module that would be observed, was decided upon by the teachers in early discussions with me. The only stipulations were that it be a Key Stage 3 class, who were going to be taught what the teachers perceived to be an environmental geography module. As described in more detail below, the observed modules focused on: Tropical Rainforests (case study 1), Energy (case study 2), and The Impact of Development (case study 3). The selection of the student participants within each case study was carried out after the teacher had identified the class to be worked with. In line with the small-scale nature of the study, I decided to work with small numbers of students (four in each case study). The principal concern underpinning this was the importance of generating rich descriptive data from the students, which I believed would be facilitated by working with a smaller number. I was also conscious of wanting to have a number which could withstand pupils being absent from a lesson, or worse still, dropping out during the process.

The selected students in each case study represented a purposive sample of four individuals who:

- Had interacted with me during my early classroom visits
- Were interested to be involved in the study
- Were not uniform in terms of gender or teacher-perceived ability.

After the first case study, I also specifically tried to avoid selecting four individuals who, in my judgement (based on classroom observation, informal interaction with the class, and discussion with the teacher), were all similarly positive and enthusiastic about their geography lessons. I was hoping instead for a mixture, so that there would be a greater chance of a variety of perspectives upon, and experiences of, lessons.

For each case study, the data generation methods included:

- Informal classroom observation and audio recording of lessons – this focused not only on gaining an audio record of each lesson, but also a written description of the teachers' and case-study students' actions and interactions and any other significant occurrences.
- Teacher and student lesson impression sheets – A4 sheets completed by the teacher and case-study students near or at the end of the observed lesson, which asked questions about their experiences of the lesson e.g. What did you enjoy in geography today? (student) and What did you feel went well in today's lesson? (teacher).
- Semi-structured student and teacher interviewing after lessons – these were characterised by trying to help the research participants to share their perspectives and experiences through creating a supportive atmosphere, grounding discussion in concrete lesson events, asking open-ended questions and then probing the meaning of their phrases and terminology, and concentrating on listening.

Analysis of the data involved detailed examination of the actions and interactions of the teacher and case-study students, particularly during those lessons where the subject matter was seen by the teacher to be issues-based as opposed to purely factual (termed 'issues lessons'). The analysis was based on lesson transcripts and observational field notes (pictures of classroom practice), as well as post-lesson interview transcripts (commentaries on classroom practice) (see Rickinson 1999b).

Fieldwork Context: Case Study 1

The observed curriculum module in case study 1 was entitled 'Tropical Rainforests' and formed part of the school's Year 8 geography course. It constituted a study of 'the characteristics and pressures of a large ecosystem – investigating how physical and human processes interact to influence the character of the vegetation'. The entire module was designed to span 12 lessons over a period of 6 weeks. This was the first time that David, the teacher, had taught this module in this school, although he had taught about tropical rainforests many times before.

An idea of the overall shape of the module can be gained from Table 1, which is a copy of the teacher's plans for the module. This also shows which lessons took place during the research period, and which lessons were analysed in detail as examples of 'issues lessons' (see shaded rows).

The school was a voluntary controlled co-educational comprehensive (1190 students aged from 11–18 years) in a small rural market town. At Key Stage 3 Geography was taught as part of humanities in mixed ability groups.

Table 1 A summary of the rainforest module in case study 1

Lessons	Questions	Activities
1 (research)	What is the rainforest like? How is it interdependent?	Watch video and draw picture to show how animals, plants, vegetation, soil and climate are related
2 (research)	How does a rainforest operate? Where are the rainforests globally, and in Brazil?	Add labels from 'How a Rainforest Works' sheet to pictures, do worksheet on Location of Rainforests
3 (research)	What factors cause rainforest climate?	Worksheet questions on Rainforest Climate
4 (research)	What are the climatic differences in Brazil? How does climate produce a rainforest?	Worksheet on 'Journey through Brazil' and 'Rainforests Climate' continued
5 (research)	Who lives in the rainforest? How do the Kayapo live?	Watch Kayapo video and do diary or poem about life as a Kayapo
6 (research)	How are the lives of the indigenous Indians changing?	Watch Emerald Forest film, completing 'story-board' whilst watching
7 (research)	Continued	Continued
8	Continued	Finish watching Emerald Forest film, discuss and write about issues in the film
9	What is the history of the Amazon rainforest? Who wants to 'develop' the forest?	Emerald Forest final re-cap, worksheet on Developing the Rainforest (before and after a dam) – colour, cut and paste, and annotate
10	How are these activities destroying the ecosystem?	Developing the Rainforest sheet continued, and worksheet on impacts of other activities e.g. cattle ranching
11	Why is the forest being developed?	Watch destroying the Rainforest video, and draw poster in pairs to show what is going on and why
12	What do the various groups of people think about this issue?	Watch video 'The World this Week' about the Altimira Convention, and complete table about different groups' perspectives e.g. Kayapo Indians, cattle ranchers etc., then prepare in groups to role play one of these perspectives
13	Continued	Class role play of the Altimira Convention
14	End of module Assessment	

Fieldwork Context: Case Study 2

The module that was focused upon in case study 2 was a 7 week Year 9 module on Energy. This encompassed a series of 14 lessons (Table 2), of which the research period spanned from lesson 6 to lesson 12. While the teacher, Alison, had taught a Year 9 Energy module before, she explained that this was 'a bit different' as she had made a few changes, namely 'introducing a bit on developing countries' and

Table 2 A summary of the energy module in case study 2

Lessons	Questions	Activities
1	What is energy? Why is it important? What types are there?	Written questions using textbook (connections)
2	How much energy, and what types, does the developing world use? What problems can this cause?	Watch video about Energy in Developing Countries, then written questions using textbook (Key Geography for GCSE)
3	What types of energy do we use in the United Kingdom? Where are different power stations located? Why?	Worksheet questions and mapping task
4	Where is Didcot? How is electricity made there? How does electricity reach us?	Worksheet exercise with local OS map
5	What environmental problems are caused by coal power stations? What is acid rain? How can it be reduced?	Written textbook questions about acid rain, then personal response to overhead transparency of acid rain-damaged trees
6 (research)	What is global warming?	Complete several questions while watching video on global warming, then write paragraph about global warming
7 (research)	What is nuclear power? What are the advantages and disadvantages of nuclear power?	Worksheet task on pros and cons of nuclear power, then writing own opinion of nuclear power
8 (research)	What are the disadvantages of nuclear power? – case study Chernobyl	Writing a paragraph about Chernobyl following teacher-led introduction, written questions on Scottish farm newspaper article, writing of opinion on nuclear power
9 (research)	What is Hydroelectric Power (HEP)? How does it work? What are the advantages and disadvantages?	Worksheet questions on HEP, followed by poster task about HEP
10 (research)	What is alternative energy? Beginning individual enquiry projects into a type of alternative energy	Teacher introduction to the enquiry projects. Gathering information from "Energy Supply" CD-ROM
11 (research)	Continued	Continued individual work using textbook and provided information sheets
12 (research)	Continued	Continued individual project work (which were completed for homework over the half Term holiday)

'removing the work on oil pollution and putting it into the Water module'. The issues lessons selected for more detailed analysis concerned nuclear power (lesson 7) and the Chernobyl disaster (lesson 8) (see shaded rows).

The school was a mixed, comprehensive upper school with 610 students aged from 13–18 years. Unlike the other two case-study schools, then, students joined this school from feeder middle schools at the beginning of Year 9, rather than from primary schools at Year 7. It was located on the outskirts of a small sized city. In Year 9 (the only Key Stage 3 year in the school), geography was taught to all students in mixed ability groups.

Fieldwork Context: Case Study 3

Case study 3 focused on a Year 9 module entitled 'One World'. It spanned 14 lessons but these were divided into two quite distinct sections: (1) Development and Interdependence (lessons 1–5); and (2) The Impact of Development (lessons 6–14). Details of the individual lessons of both of these sections are provided in Table 3, but it should be noted that all of the research undertaken within this case study occurred within the second section ('The Impact of Development'). The issues lessons

Table 3 A summary of the one world module in case study 3

Lessons	Questions	Activities
1	I: Development and Interdependence How do we show the countries of the world on a map? How do levels of economic development vary? How useful are indicators of development?	Mapping North/South divide data on world maps, undertaking work sheet on North–South differences, What is Development? worksheet
2	Continued	Continued
3	Do all countries have similar resources and opportunities? Are some countries exploited by others? What can we do about our responsibilities to other countries?	Play Oxfam's "Trading Game"
4	International trade is controlled by countries of the north – what effects does this have on people in the south? What can we do to encourage fairer trade?	Three-way carousel of group activities relating to fair trade issues e.g. looking at fairly traded products and how they are different from other products.
5	Continued	Written worksheet on world trade and fair trade issues.
6	II: The Impact of Development What is going on in the Amazon rainforest? Where is the Amazon? What is the rainforest ecosystem like?	Teacher-led introduction to human activities in Amazon using overheads, completion of world map of tropical forests, watch video about rainforests and annotate diagram of forest layers
7	What is the climate like in the rainforest? How does the ecosystem work?	Stimulus rainforest diary/story activity in groups, completing written fact-file on climate, group exercise to assemble and draw an ecosystem

(continued)

Table 3 (continued)

Lessons	Questions	Activities
8 (research)	What are the environmental effects of deforestation? What is the Carajas Project? What are my first impressions of this Project?	Teacher-led class brainstorm on the importance of tress, "Destroying the Balance" worksheet, watch brief video clip of Carajas Project and record initial impressions.
9 (research)	What and where is the Carajas Project? What effects has the project had on the environment and the local inhabitants? How do I feel about this project?	Re-watch Carajas video clip, complete written factsheet about the project, watch pro-Carajas video, record own viewpoint, watch anti-Carajas video, record own viewpoint again.
10 (research)	What different views are there on the Carajas Project?	Brief consideration in pairs of different views on Carajas (on overhead), role play preparation in groups of particular roles, and then in mixed groups
11 (research)	Continued, and How do I feel about this project?	Continued role play preparation, and then videoing of each group's performance, whole class watching of the video of all groups, final recording of own viewpoints.
12	What effect has rainforest development had upon local inhabitants?	Watch video and complete worksheet about "The Yanomami – The Dying Tribe", completing summary diagram of impacts upon people and the environment
13	End of Module Assessment	
14	Continued	

selected for more detailed analysis were those relating to the Carajas Project, lessons 9–11 (see shaded rows).

This school was a large, mixed comprehensive (2000 students aged from 11–19 years) in a rural market town. Geography was taught in mixed ability groups as part of humanities, along with successive modules of history and religious studies.

Hopwood Study

Methodological Details

Several classroom- and non classroom-based techniques were used to generate data. The former consisted of: loosely structured semi-participant classroom observation; photocopying pupils' written work; and post-lesson interviews. Two of the four non classroom-based techniques involved talking to pupils about images and how they relate to school geography. In self-directed photography pupils took photographs using disposable cameras, while images were chosen by the researcher in the supplied photo elicitation technique. Images were replaced with researcher-supplied questions in the geographical questions technique, and a concept mapping task was developed in which pupils considered links between topics they had studied. All case pupils participated in all non classroom-based techniques once, and attended between seven and nine post-lesson interviews. Further details of these techniques are given in Hopwood (2006).

The 3-year timeframe for completion of this study was a constraining factor and meant that fieldwork could not be conducted in more than three sites, especially given the requirement to undertake analysis between sites. Three different school sites brought with them different departments, schemes of work, and teachers, and thus different lesson contexts. The sampling of teachers and schools was determined by criteria set out by the researcher. That of classes reflected practical constraints with respect to timetabling and teachers' wishes and may be described as opportunistic.

Criteria used to identify suitable sites focused more on individual teachers than school-wide concerns. During further piloting, lessons led by several different teachers were observed and used as contexts for one-to-one post-lesson interviews with over thirty pupils. When lessons involved large amounts of note-taking and/or textbook work, or strongly didactic whole-class teaching, post-lesson interviews were consistently more difficult and less effective. This seemed to be because pupils appeared to have less to talk about and there was much less variation in the sorts of comments they made about such lessons.

Conversely, lessons led by teachers who typically set up a variety of experiences within a lesson and who varied the types of tasks from lesson to lesson proved conducive to stimulating lively interactions with pupils and generating rich data. The decision was thus taken to select three teachers who tended to set up lessons in this way. The department through which this research was conducted has close links with schools, and Dr. Graham Corney (primary supervisor for this study) was able to recommend suitable teachers. Initial visits and two-week immersion periods in each class ensured that the chosen classroom contexts were as required.

Had different teachers been chosen, the post-lesson interviews would have focused on different experiences and thus to an extent the choice of teachers influenced the data generated. However post-lesson interview data differed within school contexts, demonstrating that this influence was only limited. The influence of teachers and teaching on pupils' conceptions was not a focus of this study.

All research was subject to the informed consent of the teachers, Heads of Geography and Heads of School involved, to whom a proposal was sent, and from whom a written replying giving consent was obtained.

In each site a Year 9 class was selected as the basis for lesson observation and from which two case pupils were selected. Year 9 was chosen as this is the final year of compulsory geography study in England and thus pupils had the most experience of school geography before choosing subjects for GCSE.

In some cases teachers ruled out 'lower ability' or 'bottom set' classes. Although banding was not a sampling criterion the possibility that teachers might be less comfortable with researchers in their more challenging lessons was considered. Preference was given to classes whose lessons were in the mornings so that interviews could take place shortly afterwards during the lunch break. At Springfield school geography classes were not streamed, and the choice of class reflected timetabling, while at Belmont and Cedar Grove schools pupils were streamed in geography, and both the teachers and timetable influenced the choice of a 'middle' and 'top' set respectively.

Two pupils (one male, one female) were selected from the chosen class in each school in a process guided by ethical considerations and practical requirements regarding pupils' willingness and availability to participate. During a period of immersion in the classroom and through consultation with the teacher, pupils were ruled out of the study for the following reasons:

1. If they appeared ill at ease in interactions with the researcher during lessons
2. If the teacher advised that they would often not be free to attend interviews due to extra-curricular activities or the likelihood of having to attend detentions
3. If the teacher advised that they might feel pressure to give consent even if they did not want to participate.

This elimination stage followed a criterion-based sampling strategy, and resulted in a pool of around 20 eligible pupils, from which two pupils were selected at random (stratifying to ensure gender balance). Each time the first two pupils approached gave initial consent to participate.

Measures of ability or achievement did not factor in the selection of pupils in the present study, although (as shown below) these qualities did vary among the six case pupils. The aim was not to document learning, nor to identify misunderstandings or misconceptions, so notions of ability or achievement were thus less relevant. Furthermore pilot work, which involved 60 pupils from two 'mixed-ability' classes, did not suggest that pupils' ability or past achievement affect their conceptions of school geography. Rich data involving complex ideas were generated with some lower-achieving pupils, while some higher-achievers generated thinner data based around quite simplistic concepts.

The sampling of classes and the criteria used in sampling pupils may have favoured more 'able' or 'higher achieving' pupils. Information was gathered about the case pupils in this respect, after they had been picked and had agreed to participate. Lisa was among the top five in her class (both in geography and across other subjects), while Bart was described by his teacher as a 'lower-middle achiever' – as was reflected in his past grades. Both Sara and Matt were among the higher achievers in their 'middle set', achieving National Curriculum levels above average for their age. Jenie and Ryan were both selected from a 'top set' and both had attained high marks throughout Key Stage 3 in geography and other subjects, although Ryan's achievement was less consistent than Jenie's.

Fieldwork Context: Springfield School

Springfield school is located in a small market town. It is a voluntary aided comprehensive co-educational secondary (11–18) school, drawing pupils from the town and surrounding rural areas. It is the only secondary school in the town, with around 1200 pupils. Attainment of pupils on entry is average and eligibility for free school meals and ethnic diversity are very low. The 2000 Ofsted report noted above-average attainment in geography and consistently 'good' or 'very good' teaching in the department.

At the time of research the geography department consisted of three full time teachers (all with geography degrees), and one who taught geography and Physical Education (PE). The teacher involved in this research was Head of Year 9 and in his sixth year of teaching at the school (which he joined as a Newly Qualified Teacher). At Key Stage 3 pupils were taught in eight 'mixed-ability' classes of around 28 students. In the chosen class pupils sat in pairs; Lisa and Bart did not sit together. Springfield school was the site for Lessons A-H.

Lessons A and B: Population Board Game

Pupils worked on a population board game task for two lessons, having previously studied population structures in the United Kingdom and Tanzania. The task was to create a board game that should explain the population problems either country faces, or encourage changes that would alleviate those problems. The format of the game was left open for pupils to decide.

Lesson C: West Side Story

Pupils were shown a video clip of the song 'America' from West Side Story. A sheet was given out with the lyrics written out in full. Different pupils read them out, occasionally stopped by the teacher who asked questions about the story about migration being told in the song. While the clip was shown to pupils a second time, the teacher wrote instructions for four tasks, which pupils worked on for the remainder of the lesson: '(1) Briefly explain the song and say what is happening; (2) Say who likes and dislikes America; (3) Give three reasons why for each view; (4) Write what you think would happen to the migrants next'.

Lesson D: Migration Timelines

An image was projected showing a group of people ('Vicky's family') being driven away from one place and drawn to another. The teacher asked what this was showing and what it had to do with the last lesson. After a discussion of the push and pull factors which cause migration, pupils were given a sheet with a timeline showing when and why Vicky's family migrated. The teacher explained the diagram and asked pupils to construct 'their own timeline, either about their family, or about a made up person', including as many push and pull factors as possible; the location and timeframe were left open to pupils.

Lesson E: Britain's Migration History

The lesson began with a discussion of public opinion about migrants living in Britain. The teacher described the aim of the lesson as to understand why there are so many migrants in Britain. Pupils were given a comic strip showing aspects of Britain's migration history. The 'myths' that migrants are responsible for unemployment and have no right to be in Britain were contrasted on this sheet with the 'truths', for example that migrants were invited to Britain and contribute to the present standard of living. Pupils' reacted strongly (but in different ways) to the sheet, and for the remainder of the lesson the teacher managed an impromptu debate on the issue.

Lesson F: Weather and Climate

The teacher introduced a new topic – 'weather and climate' – and asked pupils what the two terms mean. He then explained the causes of global climate variations using a globe. The first task was to choose two images of different weather conditions from a selection displayed around the room and write five descriptive words about them. Pupils then swapped books and had to work out which image their partner had described. Having swapped their books back, pupils used the descriptors to write five sentences for each image.

Lesson G: Tropical Climates

Pupils were asked to write 'Characteristics of climates in the tropics' in the middle of a fresh page in their books, forming the centre of a spider diagram. A video was shown during which pupils annotated their diagram with information about tropical climates, in terms of physical conditions and how people have adapted to those conditions. The final task was to shade tropical climate areas on an outline map of the world and, using an atlas, label six countries found within that zone.

Lesson H: Global Climatic Variations

After a brief discussion of where tropical climates occur and what those areas are like (often rainforests), a video was shown, this time describing changes in climate as latitude increases towards desert regions. Afterwards pupils were asked about the characteristics of deserts and to create a spider diagram similar to that they had done for tropical climates. The second half of the lesson focused on temperate European climates, and again pupils made notes on their characteristics in the form of a spider diagram. The final task was to shade in the desert and temperate zones on the outline map of the world (from the previous lesson), again identifying six countries in each.

Fieldwork Context: Belmont School

Belmont school is one of three schools in a large market town. It is a comprehensive co-educational 11–18 school of around 800 pupils. Attainment of pupils on entry is average and both the proportion of pupils of non-white ethnic backgrounds and those eligible for free school meals are below the national average. The 2002 Ofsted report noted 'good' teaching and learning, and above average standards in geography.

At the time of fieldwork the geography department consisted of two full time teachers with geography degrees, although some lessons were taught by a member of the senior management team. The teacher involved was Head of Department, in her fifth year of teaching and her second full year at the school. Year 9 geography classes (six classes of 25 to 30 pupils) were banded into 'top', 'middle' and 'bottom' sets (a 'middle' set was chosen in this study). Pupils sat in pairs; Matt and Sara did not sit together. Belmont school was the site for Lessons I–O and Assessments A and B.

Lesson I – Hurricanes in Haiti

As part of the topic 'Tropical Storms' and addressing the question 'What is happening in Haiti?' (the location of Hurricane Jeanne at the time), pupils used an atlas to find information about Haiti. A video was then shown, telling the story of a Haitian woman, her family, and the hardships they faced. The teacher then showed a report on the BBC News website about the impacts of Hurricane Jeanne. In the plenary discussion the teacher asked pupils why the hurricane had such devastating impacts. For homework pupils were asked to bake cakes or buns for a charity sale, the proceeds of which would be sent to help people in Haiti. (Matt was absent for this lesson but 'caught up' by borrowing a friend's book.)

Lesson J – Hurricane Formation

This time asking 'How does a hurricane form?', the teacher talked through a 'recipe' for a hurricane which described the conditions and processes involved in its inception, development, and 'death' over land; pupils were asked to make notes during this time.

The task for the remainder of the lesson was to draw a cartoon strip of six boxes, explaining the life cycle of a hurricane. (Matt was absent for this lesson and covered the material in his own time.)

Assessment A – Hurricane Ivan

The teacher gave details of an end-of-topic assessment in which pupils were asked to research Hurricane Ivan and to write a report following a guide provided by the teacher: (1) introduction – what path it took, how long it lasted, wind speeds; (2) explanation of how the hurricane formed; (3) details of how people were affected by the hurricane; (4) what safety measures people take, are some places better prepared than others? Pupils were given their marked assignments back with comments from the teacher at the end of lesson K.

Lesson K – Responses to Hurricane Gloria

The final lesson on this topic asked 'How did people respond to Hurricane Gloria?'. The teacher explained that the purpose of the lesson was to classify the different responses of authorities, businesses and individuals. A newspaper report was read aloud and then pupils worked in pairs using three different colours to identify responses associated with each group. A second sheet formed the basis of the final task in which pupils worked individually, writing a paragraph about how the three groups reacted to the hurricane.

Lesson L – Types of Rainfall

This lesson was a continuation of an unobserved lesson in which a new topic ('weather and climate') was started and processes and types of rainfall studied under the key question 'How does it rain'. The first 10 minutes were spent revising what had been taught the day before. For the remainder of the lesson pupils were asked to draw a diagram of one type of rainfall and write an explanation of how it occurs underneath.

Lesson M – Rainfall in Britain

In reference to the question 'What is Britain's rainfall pattern like?' pupils were given an outline map of Britain and a textbook to help draw a choropleth map showing areas of high, medium and low precipitation. The teacher asked pupils what they noticed about the pattern and asked them to complete a series of sentences describing and explaining rainfall distribution in Britain. For the plenary task suitable locations were to be chosen for a fictional frog and cactus wanting to thrive in the United Kingdom.

Lesson N – Climate Graphs

The key question was 'Can I draw climate graphs?'. The teacher showed a climate graph using the overhead projector, and pupils read off various types of information from it. They were then asked to complete the sentence 'climate graphs show…' before sheets were distributed giving climatic data for London and Manaus (Brazil) so that pupils could draw a climate graph for each place.

Lesson O – World climates

This lesson was centred on the question 'What are the main world climates and what are their causes?'. An outline map of the world was given to pupils who were asked to mark on and identify four major climate zones and to match them with descriptions provided by the teacher. The teacher then asked what pupils thought causes different climates before giving out a second sheet showing the sun's rays 'hitting' different parts of the world in different concentrations. Once this had been explained, a video about each of the four main climates was shown, and pupils added information to their world maps describing the climate, vegetation, and people in each zone.

Assessment B – Tornadoes

The teacher set up project work which took place during two lessons and a homework slot. The lessons were in the computer room, and pupils were given a list of websites to help them find the necessary information. The task was to produce a poster, leaflet or booklet about tornadoes including the following points: a definition of a tornado; what causes them; what effects they have; pictures and diagrams; and a glossary of geographical words.

Fieldwork Context: Cedar Grove School

Cedar Grove school is in Papford, formerly a small market town, now home to two large commuter estates. It is one of two comprehensive co-educational 11–18 schools in the town, and draws 1100 pupils from the town, estates, and surrounding villages. Eligibility for free school meals and pupil attainment on entry are below average and the proportion of children with statements is above average. The 2003 Ofsted report noted 'well above average' standards in geography across the key stages, summarising the department as 'very good'.

The teacher involved in this study was Head of Department, in her seventh year of teaching and at the time the only full-time geography teacher with a geography degree. Two members of the senior management team taught some geography lessons. Year 9 geography classes were banded into 'top', 'middle' and 'bottom' sets. The 'top' set selected in this study comprised a number of pupils who were previously in 'lower' sets due to the school policy of 'promoting some pupils for behavioural reasons'. Each of the Year 9 geography sets comprised between 20 and 30 pupils. In the chosen class pupils sat in groups of between four and six; Jenie and Ryan were in different groups. Cedar Grove was the site for lessons P to X and Assessment C.

Lesson P – What Is Brazil Like?

Under the key question 'What is Brazil like?' pupils wrote a list of things they would like to find out about Brazil. After a class discussion, pupils were given an outline map of the country and the main task was to 'smother it with information' gleaned from atlases and worksheets. During the plenary pupils were asked: Is this what Brazil is really like? What problems/inaccuracies are there with this task? (Ryan was absent for this lesson.)

Lesson Q – Debate About Deforestation in Brazil

This lesson followed a homework activity in which pupils were asked to research deforestation issues in preparation for a class debate. The big picture was 'Trees or Televisions?' and the key question was 'Does deforestation mean development or disaster?'. The starter activity involved a class discussion about effective arguments and debating. Then every pupil read out a short speech for or against deforestation after which they responded and made counter-arguments. The teacher set up the plenary activity and homework in which pupils wrote the main arguments from the debate on a poster.

Lesson R – Sustainable Management of Rainforests

This lesson was entitled 'I can recognise sustainable ways to manage the tropical rainforest'. A discussion recalling points from the debate and clarifying the meaning of sustainability preceded consideration of different management strategies. Pupils read information sheets and selected one approach to summarise in their books. In the plenary pupils explained to the class how their chosen strategy worked and reasons for their choice.

Assessment C – Brazilian Rainforests

Pupils were given an assignment to complete over the vacation asking them to: explain why the Brazilian Government would want to deforest; name groups who exploit the rainforest and groups who oppose such activities and say why they do so; label a diagram illustrating the impacts of deforestation on people and the environment; outline local and worldwide arguments that arise; explain how they personally might have been involved in deforestation in Amazonia; and to describe two possible management strategies indicating who would support them and who would be affected. Finally pupils explained which they thought was the most suitable plan and how it might affect them.

Lesson S – Self Assessment of Rainforest Assignment

In this lesson pupils marked their own rainforest assignments. During the starter activity the teacher asked about what is required if a question asks you to describe, identify, explain, or predict. A guide to possible answers was given out although the teacher stressed that there might be valid alternatives. Pupils then identified two things they could do to improve their work and handed their assignments to the teacher.

Lesson T – Defining Development

The key question was 'What do we mean by development?'. In the starter activity pupils jotted down initial definitions of development. The main segment of the lesson involved pupils moving round in groups completing different activities at each table, writing what they had learned about development from each. The resources included an Oxfam appeals video, a series of photographs, sheets detailing working hours around the world, and articles about changes in Papford (the local town). The aim was to produce a more informed and refined definition of development at the end of the lesson (this formed the plenary activity).

Lesson U – Measuring Development

Under the big picture 'What is development?' and the key question 'How can we measure development?' pupils discussed in groups what development is, what others think, and why definitions differ. They then reported on their ideas and disagreements to the class. Pupils were given a sheet listing a selection of indicators used to measure development and were asked to rank them in order of importance, justifying their decisions. The main task was to produce a poster showing what development is and how it can be measured. In the plenary discussion the teacher asked about problems with measuring development.

Lesson V – The Trading Game

Instructions for a trading game were explained by the teacher. Each group of pupils represented a country and their aim was to produce as many paper shapes of particular dimensions as possible, depositing them at the 'bank' and earning money accordingly. Making such shapes required 'raw material' (paper) and 'technology' (pencils, a ruler, protractor and pair of compasses). Groups were given packs comprising 'raw materials' and 'technology' in different amounts, some finding they had ample supplies and others struggling to produce anything. In the plenary pupils recounted their experiences and the analogy with inequalities in global trading systems was emphasised.

Lesson W – Banana Trade

Within the big picture 'Trade and Development', two key questions were posed: 'Is the banana trade fair?' and 'What impact does this have on different groups of people?'. In the starter activity pupils were asked how they thought the profits from the banana trade should be divided. They were then given a printed statement from one of three banana growers. Pupils prepared a speech in which they were asked to 'put themselves in the shoes of their character and describe who they are, what they feel about the banana trade and how it affects them'. Pupils then told others on their table about their character before writing 'Bananas are our survival' as a title in their books for the final task. In this they wrote about the characters and responded to the title statement from three perspectives.

Lesson X – Fair Trade Poster Picnic

Pupils brought in Fair Trade products and were allowed to consume them in the lesson. The key question was 'Can Trade be Fairer?' and for the starter activity the teacher asked pupils what they would want to know about Fair Trade, writing their questions on a spider diagram on the board. Using a variety of resources, pupils spent the rest of the lesson working in groups producing posters about Fair Trade. During the plenary the teacher asked whether pupils would encourage their parents to buy Fair Trade products, and a range of views was offered by pupils. Homework was to explain Fair Trade to one of their parents and get them to write a written response for pupils to bring to class.

Lundholm Study

Methodological Details of Doctoral Project

Lundholm's doctoral thesis comprised three case studies, chosen with an interest in studying students' learning in environmental education in higher education, and were selected with a variation in the course *content*. This was seen as interesting since previous research had mainly focused on learning in the natural sciences, and also because possible differences in the learning process due to content could be explored. The first case study focused on engineering students following a course in ecology which drew on a range of natural sciences. The second explored biology students' experiences of a course on 'Environmental control for biologists' in which a major assignment involved 'environmental auditing' relating to the social sciences and business and administration. The third case study, on doctoral students' interpretations of environmental research and the task of writing a thesis, was chosen because it involved both the natural and social sciences.

Fieldwork Context: Case Study 1

The Royal Institute of Technology (KTH) is situated in Stockholm and is the largest institute of technology in Sweden, accommodating well over 17,000 undergraduate students. In 1999 an environmental policy and plan of action was put into effect at KTH. As a result the civil engineering programmes contain compulsory courses in either ecology, environmental science or environmental protection.

The compulsory ecology course consisted of twelve two-hour lectures, with the head teacher holding six of them, during the period of March to June 1999. Other teachers in the department held the remaining six lectures. The course also contained 'laboratory' group work and excursions. There were 100 students in the class and 91 of these took the final exam, ten of which subsequently failed. Parallel to the course in ecology the students studied general chemistry, and differential equations and transformations.

The content of the ecology course was as follows (in brief):

- Different parts of the ecosystem including soil (geology), water (hydrology), the atmosphere (meteorology), and living organisms (biology).
- The organization and function of the ecosystem, especially energy flows, the hydrological cycle and the chemical cycles.
- Ecology on an individual, population and social level, with special focus on the aspects of landscape ecology and the conservation of biological diversity.
- Human beings as part of the ecosystems and the effects of human activity.
- The use of ecological knowledge in contexts of planning and exploitation and ecological technology.

The aim of the course, as stated in the student handbook, was that the students acquire knowledge about ecological theories and principles and their connections to adjacent disciplines, and understand applications of ecological science in technology, planning and town and community building.

There were 100 students in the class, six of whom, two women and four men, were selected for interview. The interviews lasted between 50 and 70 minutes. Interview questions concerned (i) their reasons for studying an engineering programme and becoming a civil engineer, (ii) their conceptions of their first year at university and the ecology course in general, (iii) what they had learnt and (iv) how they related the course content to their future profession.

Fieldwork Context: Case Study 2

In the second case study observations were made of the lectures in a supplementary course at Stockholm University in 1998. Entitled 'Environmental control for biologists', the course was multi-disciplinary in character, adopting societal perspectives on environmental issues with lectures on politics, economy, environment and technology, environmental law and international environmental control. At the end of the course the students were offered five different themes to work with. In the case study reported, a group of four students, three women and one man, chose to work on a task on 'environmental auditing' and the use of 'environmental reports' in business. The students' group work was observed, hand-written notes were made of their activities and the discussions were tape-recorded. The work progressed over a three-week period.

Fieldwork Context: Case Study 3

The third case study was conducted in 1999, also at KTH. Six postgraduate students, four women and two men, were interviewed concerning their environmental research, choice of topic and interest in pursuing postgraduate studies. Three of the students had just begun, or were in the process of planning for their postgraduate studies and the other three intended to present their licentiate thesis within the coming year. The interviews lasted between 60 and 90 minutes. All of the interviews were transcribed in full.

Methodological Details of Postdoctoral Project

Also incorporated into this book are findings from a related postdoctoral project (2004–2007) which aimed at furthering understanding of the ways values play a role in learning and conceptual development. Based on previous findings showing how engineering students' were challenged by different views on nature (anthropocentric and ecocentric) when studying ecology, the assumption was made that economics students could equally find learning about the environment and ecology challenging due to their values and beliefs.

Interviews were conducted with students in a masters course on 'Sustainable Enterprising' with a focus on environmental business and management at Stockholm University. These students were of interest due to their mixed educational background in biology and economics. They were either studying for degrees in economics or in biology and as part of their programmes took smaller component courses in biology/ecology or economics respectively. Within the group of 11 students, two had degrees in economics, and among the remaining nine students, four had studied at different inter-disciplinary environmental programmes (environmental communication, ecological economics, environmental and sustainability) and five had a degree in biology followed by courses in economics. Thus the group comprised of students who had first studied economics and then ecology, and vice versa.

All the students were interviewed individually. Interviews lasted between 45 to 60 minutes and were tape-recorded and transcribed in full. The following questions provided a general rubric for these interactions:

- Why have you chosen this masters course on Sustainable Enterprising?
- What is your professional and educational background?
- If you have a degree in biology or environmental science, what was your experience when entering the subject of economics?
- How do you perceive the relation between economy and environment?

Appendix II: Development of the Lenses

This appendix describes in greater detail the processes through which the three lenses used in this book were developed.

Introduction

The process described here began in 2004 when Lundholm and Rickinson first noticed similar issues emerging in their separate studies. These complementary findings have been reported elsewhere (Lundholm and Rickinson 2005, 2006; Rickinson and Lundholm 2008). Our empirical scope was soon widened to incorporate Hopwood's work (Hopwood 2006, 2007a, b, c, 2008, 2009).

Despite our excitement in looking across our separate studies to find common outcomes, we shared dissatisfaction with the mere reporting of complementary findings on students' experiences and learning. It struck us that there was more that could be said, and learned, from our studies. We felt instinctively that we each had something different to say about the three studies and what we could learn from them.

Our original studies reflected different key research interests in (1) emotions and values in learning (Lundholm), (2) students' conceptions of relevance in environmental education (Hopwood), (3) difference in viewpoints and relations between teachers and students in environmental education (Rickinson). These distinct areas of interest formed the starting points in the development of each lens, and continue to shape them. However the lenses have come to represent, and be informed by, much more than our original lines of inquiry and modes of questioning and analysis, as we demonstrate below.

Empirical Focus

The first step in translating an individual interest into a useful conceptual device was to share our raw data so that we could review (question, analyse, interpret, make sense of) each other's transcripts in our own distinct ways. This process involved exploring similarities and differences between data and findings, and meeting to discuss emerging ideas. We conceived our ways of approaching the material as looking at the same thing (students' learning and experiences) through different lenses: the metaphor represents ideas of bringing certain things into focus, offering different but equally valid ways of seeing something.

Over time the way we each looked at, interpreted and understood our own data changed as a result of engaging with the empirical material from the other studies. For example, Lundholm's original work focused on emotions and values as relating to the learner on a personal level, within an affective domain influencing student cognition and learning. The related lens is, however, broader, and also explores learners' views on values and opinions as part of subject matter. Lundholm has thus come to see new things in the empirical material she already understood in depth, and now sees traces of learners grappling with how to learn about values and opinions.

Engaging Other Research on Environmental Learning

We were mindful that we had to attend both to processes of interpreting our shared data, and to the process of developing robust and useful conceptual tools. Indeed, these were mutually reinforcing, and neither could proceed without the other. An important means to refine each lens involved looking beyond our shared data to other research on environmental learning (remaining within our boundary of formal learning contexts). Looking at this wider corpus of studies through each lens enabled us to better understand where each applied, what it revealed, how it could be defined, what the important ideas and concepts were. As well as providing a mechanism to assist in conceptual development, this process also demonstrated their relevance and utility in making sense of a much larger body of research.

This can be exemplified in the context of the lens which looks at/for issues to do with relevance. In research explicitly aiming to understand the process or experience of environmental education, we found numerous references to issues of relevance. These included the reported desire among Australian learners for practicality and relevance in environmental learning (Connell et al. 1999), and the potential of (perceived) relevance of environmental learning as a means to address student disaffection in English schools (Battersby 1999). Engaging with such studies helped gain a broader sense of what relevance might mean in the context of environmental learning, and also helped us gather together a number of studies to which the lens could be applied. DiEnno and Hilton (2005) explored US high school students' interest in and attitudes towards particular (environmental) content, and Smith-Sebasto and Walker (2005) discuss learner reactions to environmental learning with reference to the affective domain. Studies such as these helped us clarify what the lens did and did not seek to explore, what concepts were most appropriate (e.g. relevance may relate to but does not equate to interest or attitudes).

Engaging Wider Research and Theory

The conceptual clarity and power of the lenses were also enhanced by looking beyond empirical studies of environmental learning to literature of a more theoretical nature which was often related to other areas of learning (e.g. science, history) or generic (i.e. relating to learning in general) rather than specific to environmental education. The key ideas related to all three lenses are discussed in the chapter titled "Lenses for Understanding Environmental Learning", but we provide a more detailed example here with reference to the lens which looks at/for emotions and values.

Emotions have gained increasing attention in recent years, and have been investigated in different ways, as in students' anxiety in test situations, or in relation to metacognitive skills for dealing with learning tasks (Efklides and Volet 2005). The emotional aspect of learning also relates to motivation and students' engagement in schoolwork and instruction. With regard to the latter, there has been an increasing interest in understanding the ways motivation, emotions and values are an important part of the process of conceptual development and change (Pintrich et al. 1993; Watts and Alsop 1997; Sinatra and Pintrich 2003, Sinatra 2005). Scholars within this field, such as Watts and Alsop (1997), address this need in contrast to the model presented by Posner et al. (1982), which they find looks at the student and the process of learning as being too rational. The work by Claxton (1989) helped us see how we might conceptualise what learners are trying to do when looking at/for emotions and values in environmental learning:

> What they are 'up to' (in all senses) is the outcome of a tacit decision-making process based only on subjective estimates of competing priorities, opportunities, demands, resources and risks, which is, in their terms, sensible and vital. Only if the emotional/motivational factors in this decision permit or encourage intellectual learning and the subjective assessments are accurate, will achievement be limited by such cognitive factors as 'ability' or alternative conceptions. (Claxton 1989, p. 159)

Such literature helped us refine dimensions of the lens relating to questions about learners' emotions and values as part of the learning process. However the initial re-viewing of data across our three studies had pointed towards values and emotions as significant in another way: relating to learners' conceptions of values in subjects and subject matter. Our understanding of how we might look at/for values and emotions in environmental learning was informed by work on scientific literacy, which has explored learning about issues that link natural science and society (including global climate change). Scientific literacy has been discussed as an important aspect of enabling learners to take part in public debate or make justifiable decisions on scientific – often environmental – issues presented in media (Driver et al. 1996). We were able to situate what we had noticed in our original data as learners' reactions to values-rich subject matter in the context of work on scientific literacy, and in studies in which learners were found to distinguish the subjectivity of opinion with the objectivity of scientific knowledge (Zeidler et al. 2002).

Reflections on the Process

The way we developed the three lenses is best thought of in terms of a number of inter-related processes rather than a step-by-step approach, because there was much iteration and oscillation between processes. This said, we had a clear starting point in the sharing of each other's raw data, and we would suggest this to be a fruitful beginning if others were to attempt a similar process in the future. Engaging with other empirical material, and re-engaging with your own, are complemented and supported by engaging with other studies (from within a particular bounded context – in our case environmental learning in formal contexts), and with literature from less directly related areas (as with reference to conceptual change and scientific literacy in the case of one of our lenses). One aspect that must be stressed is the collaborative nature of the exercise – regular meetings were crucial to help us understand each other's empirical material, verify interpretations, ask critical questions, point to relevant literature, and so on. The process we describe is one which inherently involves more than one person, so that the looking across data, other research, and theoretical literature is complemented by a variety of in-person perspectives.

Our work resulted in three distinct lenses being developed in the context of environmental learning in formal settings. We believe strongly that others might be created which make sense of formalised environmental learning in different – equally valid, useful, insightful – ways, and we would encourage others to build on the lines of enquiry and questioning they are familiar with to develop new lenses which can be used to look across and make sense of collected empirical work in other ways. We also acknowledge that our focus on formal settings excludes many less formal contexts in which environmental learning takes place. It may be that the lenses we have developed offer some purchase on a range of other learning settings – but we would argue that an iterative process of engaging with different empirical and theoretical fields would be likely to generate new lenses which look at or for other things in different ways. The power of the lenses comes from drawing on concepts and being informed by relevant scholarship and their application to contexts other than those in which they were developed. However we can imagine families of lenses which explore similar issues but which draw on the concepts and ideas most relevant to their context.

We now consider conditions that facilitated the processes described here, and in doing so point to some limitations in the scope or potential of similar work being repeated in different contexts. On a practical level, the ability to share and understand each other's raw data was crucial – we needed a common language (Lundholm's transcripts were translated from Swedish to English), and to furnish each other with sufficient methodological detail so we each knew how the data came about. For ethical reasons it was important to ensure any shared data were suitably anonymised. Furthermore, we have found the development of lenses to require considerable time in individual work as well as for face-to-face meetings. The multi-faceted, iterative, and collaborative nature of this type of work does not lend itself to speedy outcomes; indeed a slowness of pace may even be necessary to allow the full richness of the lenses and the range of literature (empirical and theoretical) to which they relate to be identified.

References

Alexander P, Schallert D, Reynolds R (forthcoming) What is learning anyway? A topographical perspective considered. Educ Psychol

Ballantyne R, Packer J (2002) Nature-based excursions: school students' perceptions of learning in natural environments. Int Res Geogr Environ Educ 11(3):218–236

Ballantyne R, Packer J (2005) Promoting environmentally sustainable attitudes and behaviour through free-choice learning experiences: what is the state of the game? Environ Educ Res 11(3):281–295

Ballantyne R, Connell S, Fien J (1998) Factors contributing to intergenerational communication regarding environmental programs: preliminary research findings. Aust J Environ Educ 14:1–10

Banks F, McCormick B (2005) A case study of the inter-relationship between science and technology: England 1984–2004. Paper presented at the 15th International Pupil Attitudes Towards Technology (PATT) Conference, Haarlem, the Netherlands, 18–22 April

Barratt R, Barratt Hacking E (2008) A clash of worlds: children talking about their community experience in relation to the school curriculum. In: Reid A, Bruun Jensen B, Nikel J, Simovska V (eds) Participation and learning: perspectives on education and the environment, health and sustainability. Springer, London

Battersby J (1999) Does environmental education have 'street credibility' and the potential to reduce pupil disaffection within and beyond their school curriculum? Cambridge J Educ 29(3):447–459

Beane JA (1997) Curriculum integration: designing the core of democratic education. Teachers College Press, New York

Biddulph M, Adey K (2004) Pupil perceptions of effective teaching and subject relevance in history and geography at key stage 3. Res Educ 71:1–8

Bixler RD, Floyd MF (1999) Hands on or hands off? Disgust sensitivity and preference for environmental education activities. J Environ Educ 30(3):4–11

Bogner FX (1998) The influence of short-term outdoor ecology education on long-term variables of environmental perspective. J Environ Educ 29(4):17–29

Bonnett M, Williams J (1998) Environmental education and primary children's attitudes towards nature and the environment. Cambridge J Educ 28(2):159–174

Boyes E, Stanisstreet M (1994) The ideas of secondary school children concerning ozone layer damage. Global Environ Change 4(4):311–324

Boyes E, Stanisstreet M (1996) Threats to the global atmospheric environment: the extent of pupil understanding. Int Res Geogr Environ Educ 5(3):186–195

Boyes E, Stanisstreet M (1997) Children's models of understanding of two major global environmental issues (ozone layer and greenhouse effect). Res Sci Technol Educ 15(1):19–28

Brandt CB (2007) Using landscape to explore the relationships between epistemology and place in science. Paper presented at the Annual Meeting of the American Educational Research Association, Chicago, 9–13 April

Brody M, Tomkiewicz W (2002) Park visitors' understandings, values and beliefs related to their experience at Midway Geyser Basin, Yellowstone National Park, USA. Int J Sci Educ 24(11):1119–1141

Brook A, Briggs H, Driver R (1984) Aspects of secondary students' understanding of the particulate nature of matter. Children's Learning in Science Project, Centre for Studies in Science and Mathematics Education, University of Leeds, Leeds

Brooks C (2003) Investigating the geography behind the news. Teach Geogr 28(2):70–73

Bruner J (1960) The process of education. Harvard University Press, Cambridge, MA

Buehl MM, Alexander PA (2005) Motivation and performance differences in students' domain-specific epistemological belief profiles. Am Educ Res J 42(4):697–726

Chan KKW (1996) Environmental attitudes and behaviour of secondary school students in Hong Kong. The Environmentalist 16(4):297–306

Claxton G (1989) Educating the inquiring mind. Harvester Wheatsheaf, London

Claxton G (1991) Cognition doesn't matter if you're scared, depressed or bored. In: Adey P (ed) Adolescent development and school science. Falmer, London

Cobb P, Confrey J, Disessa A, Lehrer R, Schauble L (2003) Design experiments in educational research. Educ Res 32(1):9–13

Connell S, Fien J, Sykes H, Yencken D (1998) Young people and the environment in Australia: beliefs, knowledge, commitment and educational implications. Aust J Environ Educ 14:39–48

Connell S, Fien J, Lee J, Sykes H, Yencken D (1999) 'If it doesn't directly affect you, you don't think about it': a qualitative study of young people's environmental attitudes in two Australian cities. Environ Educ Res 5(1):95–113

Cooper P, McIntyre D (1993) Commonality in teachers' and pupils' perceptions of effective classroom learning. Br J Educ Psychol 63:381–399

Cooper P, McIntyre D (1996) Effective teaching and learning: teachers' and students' perspectives. Open University Press, Buckingham

Corney G (2000) Student geography teachers' pre-conceptions about teaching environmental topics. Environ Educ Res 6(4):313–329

Cotton DRE (2006a) Implementing curriculum guidance on environmental education: the importance of teachers' beliefs. J Curriculum Stud 38(1):67–83

Cotton DRE (2006b) Teaching controversial environmental issues: neutrality and balance in the reality of the classroom. Educ Res 48(2):223–241

Cullingford C (1994) The attitudes of children to their environment. Cambridge J Educ 24(1):7–19

Dettmann-Easler D, Pease JL (1999) Evaluating the effectiveness of residential environmental education programs in fostering positive attitudes toward wildlife. J Environ Educ 31(1):33–39

DfEE/QCA (1999) Geography: The National Curriculum for England, Key Stages 1–3. HMSO, London

DiEnno CM, Hilton SC (2005) High school students' knowledge, attitudes and levels of enjoyment of an environmental education unit on nonnative plants. J Environ Educ 37(1):13–25

Dillon J (2003) On learners and learning in environmental education: missing theories and ignored communities. Environ Educ Res 9(2):215–226

Dole JA, Sinatra GM (1998) Reconceptualizing change in the cognitive construction of knowledge. Educ Psychol 33(2/3):109–128

Dowgill P (1998) Pupils' conceptions of learning geography under the National Curriculum. Unpublished Ph.D. thesis, University of London

Doyle W (1986) Content representations in teachers' definitions of academic work. J Curriculum Stud 18(4):365–379

Driver R (1983) The pupil as scientist?. Open University Press, Milton Keynes

Driver R, Leach J, Millar R, Scott P (1996) Young people's images of science. Open University Press, Buckingham

Edwards D, Mercer N (1987) Common knowledge: the development of understanding in the classroom. Routledge, London

Efklides A, Volet S (2005) Emotional experiences during learning: multiple, situated and dynamic. Learn Instr 15(5):377–380

Elliot J (1995) Reconstructing the environmental education curriculum: teachers' perspectives. In: OECD (ed) Environmental learning for the 21st century. OECD, Paris

Emmons KM (1997) Perceptions of the environment while exploring the outdoors: a case study in Belize. Environ Educ Res 3(3):327–344

Entwistle N, Smith C (2002) Personal understanding and target understanding: mapping influences on the outcomes of learning. Br J Educ Psychol 72:321–342

Erickson F, Shultz J (1992) Students' experience of the curriculum. In: Jackson PW (ed) Handbook of research on curriculum. Macmillan, New York

Ertl H, Hayward G, Wright S, Edwards A, Lunt I, Mills D & Yu K (2007) The student learning experience in higher education. Literature review report for the Higher Education Academy. Accessed 10 November 2008 from http://www.heacademy.ac.uk/projects/detail/lr_2007_hayward

Estabrooks CA (2001) Research utilization and qualitative research. In: Morse JM, Sawnson JM, Kuzel AJ (eds) The nature of qualitative evidence. Sage, London

European Commission (2007) Towards more knowledge-based policy and practice in education and training. European Commission, Brussels

Falk JH (2005) Free-choice environmental learning: framing the discussion. Environ Educ Res 11(3):265–280

Farmer AJ, Wott JA (1995) Field trips and follow-up activities: fourth graders in a public garden. J Environ Educ 27(1):33–35

Fien J (1992) Education for the environment: a critical ethnography. Ph.D. dissertation, University of Queensland

Fourth International Conference on Environmental Education (2007) Moving forward from Ahmebadad ... environmental education in the 21st century. Accessed 28 November 2008 from http://www.tbilisiplus30.org/Final%20Recommendations.pdf

Gayford C (1995) Science education and sustainability: a case-study in discussion-based learning. Res Sci Technol Educ 13(2):135–45

Gayford C (2004) A model for planning and evaluation of aspects of education for sustainable development for students training to teach science in primary schools. Environ Educ Res 10(2):255–271

Gee JP (2001) Identity as an analytic lens for research in education. Rev Res Educ 25:99–125

Goodlad J & Associates (1979) Curriculum inquiry: the study of curriculum practice. McGraw-Hill, New York, NY

Grace MM, Ratcliffe M (2002) The science and values that young people draw upon to make decision about biological conservation issues. Int J Sci Educ 24(11):1157–1169

Gutierrez de White T, Jacobson SK (1994) Evaluating conservation education programs at a South American zoo. J Envir Educ 25(4):18–22

Halldén O (1988) Alternative frameworks and the concept of task: cognitive constraints in pupils' interpretations of teachers' assignments. Scand J Educ Res 32:123–140

Halldén O, Haglund L, Strömdahl H (2007) Conceptions and contexts: on the interpretation of interview and observational data. Educ Psychol 42(1):25–40

Hart P (2007) Desires and resistances as drivers and barriers to environmental learning and sustainability: a Canadian perspective. In: Bjorneloo I, Nyberg E (eds) Drivers and barriers for implementing learning for sustainable development in pre-school through upper secondary and teacher education. ESD in action technical paper No 4. UNESCO, Paris

Heimlich JE (2005) Editorial. Environ Educ Res 11(3):261–263

Hopwood N (2004) Pupils' conceptions of geography: towards an improved understanding. Int Res Geogr Environ Educ 13(4):348–361

Hopwood N (2006) Pupils' conceptions of school geography: a classroom based investigation. DPhil dissertation, University of Oxford

Hopwood N (2007a) Environmental education: pupils' perspectives on classroom experience. Environ Educ Res 13(4):453–465

Hopwood N (2007b) Pupils' conceptions of geography: issues for debate. In: Halocha J, Powell A (eds) Conceptualising geographical education. International Geographical Union Commission on Geographical Education/Institute of Education, London

Hopwood N (2007c) Researcher roles in a school-based ethnography. In: Walford G (ed) Studies in educational ethnography, Volume 12: methodological developments in ethnography. Elsevier, Oxford

Hopwood N (2008) Values in geographic education: the challenge of attending to learners' perspectives. Oxf Rev Educ 34(5):589–608

Hopwood N (2009) UK high school pupils' conceptions of geography: research findings and methodological implications. Int Res Geogr Environ Educ 18(3):1–13

Hopwood N, Courtley-Green C, Chambers T (2005) Year 9 students' conceptions of geography. Teach Geogr 30(2):91–93

Huckle J (1994) Environmental education and the National Curriculum in England and Wales. Int Res Geogr Environ Educ 2(1):101–104

Hutchinson F (1997) Our children's futures: are there lessons for environmental educators? Environ Educ Res 3(2):189–201

Jensen B, Schnack K (1997) The action competence approach in environmental education. Environ Educ Res 3(3):163–179

Jovchelovitch S (2007) Knowledge in context: representations, community and culture. Routledge, London

Kaplan S (1985) Cognition and affect in environmental learning. Child Environ Q 2(3):19–21

Knamiller G (1983) Environmental education for relevance in developing countries. Environ 3(3):173–179

Kuhlemeier H, Van den Bergh H, Lagerweij N (1999) Environmental knowledge, attitudes, and behavior in Dutch secondary education. J Environ Educ 30(2):4–14

Kwan T, Miles J (1998) In the words of children and young people: the opinions and concerns about their environments of some Brisbane school students. Aust J Environ Educ 14:11–18

Kyburz-Graber R (1999) Environmental education as critical education: how teachers and students handle the challenge. Cambridge J Educ 29(3):415–432

Lai KC (1999) Freedom to learn: a study of the experiences of secondary school teachers and students in a geography field trip. Int Res Geogr Environ Educ 8(3):239–255

Lam C-C, Lai E (2003) 'What is geography?' In the eyes of junior secondary students in Hong Kong. Int Res Geogr Environ Educ 12(3):199–218

Leach J, Millar R, Ryder J, Séré MG (2000) Epistemological understanding in science learning: the consistency of representations across contexts. Learn Instr 10(6):497–527

Leeming FC, Porter BE, Dwyer WO, Cobern MK, Oliver DP (1997) Effects of participation in class activities on children's environmental attitudes and knowledge. J Environ Educ 28(2):33–42

Littledyke M (2004) Primary children's views on science and environmental issues: examples of environmental cognitive and moral development. Environ Educ Res 10(2):217–234

Lord P (2001) Pupils' experiences and perspectives of the National Curriculum: updating the research review 2000–2001. QCA, London. http://www.qca.org.uk/4831.html. Accessed 6 August

Lord P (2002) Pupils' experiences and perspectives of the National Curriculum: updating the research review 2001–2002. QCA, London. http://www.qca.org.uk/4834.html. Accessed 6 August

Lord P (2003) Pupils' experiences and perspectives of the National Curriculum: updating the research review 2002–2003. QCA, London. http://www.qca.org.uk/4836.html. Accessed 6 August

Lord P, Harland J (2000) Pupils' experiences and perspectives of the National Curriculum: research review. QCA, London. http://www.qca.org.uk/4828.html. Accessed 6 August

Lord P, Jones M (2006) Pupils' experiences and perspectives of the National Curriculum and assessment: research review. QCA, London. http://www.nfer.ac.uk/research-areas/pims-data/outlines/pupils-experiences-and-perspectives-of-the-national-curriculum-and-assessment-fourth-update.cfm. Accessed 13 October 2008

Lundholm C (2003) Att lära om miljö. Forskar- och högskolestuderandes tolkningar av ett miljöinnehåll i utbildningen/ Learning about environmental issues. Undergraduate and postgraduate students' interpretations of environmental content. Ph.D. dissertation, Stockholm University

Lundholm C (2004a) Case studies – exploring students' meaning and elaborating learning theories. Environ Educ Res 10(1):115–124

Lundholm C (2004b) Learning about environmental issues in engineering programmes. A case study of first-year civil engineering students' contextualisations of an ecology course. Int J Sustainability High Educ 5(3):295–307

Lundholm C (2005) Learning about environmental issues: undergraduate and postgraduate students' interpretations of environmental content. Int J Sustainability High Educ 6(3):242–253

Lundholm C (2007) Pricing nature at what price? A study on undergraduate students' conceptions of economics. S Afr J Environ Educ 24:126–140

Lundholm C (2008a) Discourse, cause and change: a study on economics students' conceptions of child labour. In: Öhman J (ed) Ethics and democracy in education for sustainable development: contributions from Swedish research. Liber, Stockholm

Lundholm C (2008b) Contextualisation and learning in economics: an intentional perspective on the role of values. Paper presented at the 6th International Conference on Conceptual Change, European Association for Research in Learning and Instruction, Turku, 22–25 August

Lundholm C, Rickinson M (2005) Exploring students' learning in environmental education. Paper presented at the Biennial Meeting of the European Association for Research on Learning and Instruction, Nikosia, 23–27 August

Lundholm C, Rickinson M (2006) I didn't really feel sorry for the trees, it was just something to write: exploring students' learning in environmental education. Paper presented at the Annual Meeting of the American Educational Research Association, San Francisco, 7–11 April

Lundholm C, Hopwood N, Rickinson M (2008) Developing lenses for understanding environmental learning. Paper presented at the Annual Meeting of the American Educational Research Association, New York, 24–28 March

Lynch K (2003) Editorial: children's geographies versus adult curricula. Geography 88(3):161–163

Marton F (1994) Phenomenography. In: Husen T, Postlethwaite TN (eds) The international encyclopedia of education. Pergamon, Oxford

Mason L (2003) Personal epistemologies and intentional conceptual change. In: Sinatra G, Pintrich P (eds) Intentional conceptual change. Lawrence Erlbaum Associates, New Jersey

Mason L, Santi M (1998) Discussing the greenhouse effect: children's collaborative discourse reasoning and conceptual change. Environ Educ Res 4(1):67–85

McIntyre D, Peddar D, Rudduck J (2005) Pupil voice: comfortable and uncomfortable learnings for teachers. Res Pap Educ 20(2):149–168

McKenzie M (2006) Three portraits of resistance: the (un)making of Canadian students. Canad J Educ 29(1):199–222

Milton B, Cleveland E, Bennett-Gates D (1995) Changing perceptions of nature, self, and others: a report on a Park/School Program. J Environ Educ 26(3):32–39

Molander B-O (1997) Joint discourses and disjointed courses: a study on learning in upper secondary school. Unpublished Ph.D. thesis, Stockholm University

Morgan J, Lambert D (2005) Geography: teaching school subjects 11–19. Routledge, London

Myers M (2006) Environmental learning: reflections on practice, research and theory. Environ Educ Res 12(3/4):459–470

Nagel NG (1996) Learning through real-world problem solving: the power of integrative teaching. Corwin, Thousand Oaks, CA

Nagel M (2004) Lend them an ear: the significance of listening to children's experiences of environmental education. Int Res Geogr Environ Educ 13(2):115–127

Nagel M, Lidstone J (2008) Green doesn't always mean go! Children's conceptions of environmental education. VDM Verlag Dr Müller Aktiengesellschaft & Co KG, Saarbrucken

Nespor J (1987) Academic tasks in a high school English class. Curriculum Inq 17(2):202–228

Norman M, Harrison L (2004) Year 9 students' perceptions of school geography. Teach Geogr 29(1):11–15

Nutley SM, Walter I, Davies HTO (2007) Using evidence. How research can inform public services. Policy, Bristol

OECD (1995) Environmental learning for the 21st century. OECD, Paris

OECD (2007) Evidence in education. Linking research and policy. OECD, Paris

OECD (2008) OECD workshop on education for sustainable development, Paris, September 2008. http://www.oecd.org/dataoecd/52/1/41372200.pdf. Accessed 10 November 2008

Österlind K (2005) Concept formation in environmental education: 14-year-olds' work on the intensified greenhouse effect and the depletion of the ozone layer. Int J Sci Educ 27(8):891–908

Österlind K (2006) Emotionella faktorers roll i begreppsbildning/The role of emotional factors in students' concept formation. Research report no 73, Department of Education, Stockholm University, Stockholm

Palmer JA (1995) Environmental thinking in the early years: understanding and misunderstanding of concepts related to waste management. Environ Educ Res 1(1):35–45

Palmer JA (1998) Environmental education in the 21st century: theory, practice, progress and promise. Routledge, London

Payne P (1998) Children's conceptions of nature. Aust J Environ Educ 14:19–26

Pekrun R (2005) Progress and open problems in educational emotion research. Learn Instr 15(5):497–506

Pintrich PR, Marx RW, Boyle RA (1993) Beyond cold conceptual change: the role of motivational beliefs and classroom contextual factors in the process of conceptual change. Rev Educ Res 63(2):167–199

Pollard A (2003) Learning through life—higher education and the lifecourse. In: Watson D, Slowey M (eds) Higher education and the lifecourse. SRHE, Buckingham

Posner G, Strike K, Hewson P, Gertzog W (1982) Accommodation of a scientific conception: toward a theory of conceptual change. Sci Educ 66(2):211–227

Postman N, Weingartner C (1969) Teaching as a subversive activity. Dell, New York

QCA (2007) Qualifications and Curriculum Authority website: Geography key stage 3. http://www.qca.org.uk/qca_12212.aspx. Accessed 30 August 2007

Randler C, Ilg A, Kern J (2005) Cognitive and emotional evaluation of an amphibian conservation program for elementary school students. J Environ Educ 37(1):43–52

Reid A, Nikel J (2008) Differentiating and evaluating conceptions and examples of participation in environment-related learning. In: Reid A, Jensen BB, Nikel J, Simovska V (eds) Participation and learning: perspectives on education and the environment, health and sustainability. Springer, Dortrecht

Reid A, Scott W (2006a) Researching education and the environment: an introduction. Environ Educ Res 12(3/4):239–246

Reid A, Scott W (2006b) Researching education and the environment: retrospect and prospect. Environ Educ Res 12(3/4):571–587

Reid A, Jensen BB, Nikel J, Simovska V (eds) (2008) Participation and learning: perspectives on education and the environment, health and sustainability. Springer, Dortrecht

Rickinson M (1999a) The teaching and learning of environmental issues through geography: a classroom-based study. D.Phil. dissertation, University of Oxford

Rickinson M (1999b) People-environment issues in the geography classroom: towards an understanding of students' experiences. Int Res Geogr Environ Educ 8(2):120–139

Rickinson M (2001) Learners and learning in environmental education: a critical review of evidence. Environ Educ Res 7(3):207–317

Rickinson M (2006) Researching and understanding environmental learning: hopes for the next ten years. Environ Educ Res 12(3/4):445–458

Rickinson M, Lundholm C (2008) Exploring students' learning in environmental education. Cambridge J Educ 38(3):341–353

Rickinson M, Reid A (2003) What's the use of research in environmental education? Paper presented at the Annual Meeting of the American Educational Research Association, Chicago

Rickinson M, Robinson L (1999) Environmental education research in the classroom: a shared methodological reflection by the teacher and the researcher. Environ Educ Res 5(1):77–93

Rickinson M, Dillon J, Teamey K, Morris M, Choi M, Sanders D, Benefield P (2004) A review of research on outdoor learning. Field Studies Council, Shrewsbury

Roper Starch Worldwide (1994) Environmental attitudes and behaviors of American youth with an emphasis on youth from disadvantaged areas (ED 381 599). National Environmental Education and Training Foundation, Washington, DC

Rosiek J (2003) Emotional scaffolding: an exploration of the teacher knowledge at the intersection of student emotion and the subject matter. J Teach Educ 54(5):399–412

Sadler TD, Donnelly LA (2006) Socioscientific argumentation: the effect of content knowledge and morality. Int J Sci Educ 28(12):1463–1488

Sansone C, Thoman DB (2005) Does how we feel affect what we learn? Some answers and new questions. Learn Instr 15(5):507–515

Sauvé L (2005) Currents in environmental education: mapping a complex and evolving pedagogical field. Canad J Environ Educ 10:11–37

Schindler G (1993) The Environment and School Initiatives (ENSI). II: the conflict. Educ Action Res 1(3):457–469

Schnack K (2008) Participation, education, and democracy: implications for environmental education, health education, and education for sustainable development. In: Reid A, Bruun Jensen B, Nikel J, Simovska V (eds) Participation and learning: perspectives of education and the environment, health and sustainability. Springer, London

Schoenfeld AH (2004) Multiple learning communities: students, teachers, instructional designers, and researchers. J Curriculum Stud 36(2):237–255

Schommer M (1990) Effect and beliefs about the nature of knowledge and comprehension. J Educ Psychol 82(3):498–504

Schommer M (1993) Epistemological development and academic performance among secondary students. J Educ Psychol 85(3):406–411

School Council/Nuffield Humanities Project (1970) The Humanities Project: an introduction. Heinemann, London

Scott W, Gough S (2003) Sustainable development and learning: framing the issues. Routledge, London

Shanahan M, Meyer JHF (2006) The troublesome nature of a threshold concept in economics. In: Meyer J, Land R (eds) Overcoming barriers to student understanding: threshold concepts and troublesome knowledge. Routledge, London

Shapiro BL (1994) What children bring to light: a constructivist perspective on children's learning in science. Teachers College Press, London

Simmons DA (1994) Urban children's preferences for nature: lessons for environmental education. Child Environ 11(3):194–203

Sinatra G (2005) The 'warming trend' in conceptual change research: the legacy of Paul R. Pintrich. Educ Psychol 42(2):107–115

Sinatra G, Pintrich P (eds) (2003) Intentional conceptual change. Lawrence Erlbaum Associates, New Jersey

Smith-Sebasto NJ, Walker LM (2005) Toward a grounded theory for residential environmental education: a case study of the New Jersey school of conservation. J Environ Educ 37(1):27–42

Songer NB, Linn MC (1991) How do students' views of science influence knowledge integration? J Res Sci Teach 28(9):761–784

Southerland S, Sinatra G (2003) Learning about biological evolution: a special case of intentional conceptual change. In: Sinatra G, Pintrich P (eds) Intentional conceptual change. Lawrence Erlbaum Associates, New Jersey

Stanton G (2006) D&R programmes: concepts and practice. NERF Working Paper 5.6. National Educational Research Forum (NERF): London

Sterling S (2001) Sustainable education: revisioning learning and change. Green Books, Dartington

Sterling S (2004) The learning of ecology, or the ecology of learning? In: Scott W, Gough S (eds) Key issues in sustainable development and learning: a critical review. RoutledgeFalmer, London

Stevenson R, Dillon J (forthcoming) Engaging environmental education: learning, culture and agency. Sense, Rotterdam

Stradling R, Noctor M, Baines B (1984) Teaching controversial issues. Edward Arnold, London

Talbot P (2002) Geography... more than a job. Teach Geogr 27(3):140–142

TG-C Ivy, Lee CK-E, Chuan GK (1998) A survey of environmental knowledge, attitudes and behaviour of students in Singapore. Int Res Geogr Envir Educ 7(3):181–202

Thagard P, Zhu R (2003) Acupuncture, incommensurability, and conceptual change. In: Sinatra G, Pintrich P (eds) Intentional conceptual change. Lawrence Erlbaum Associates, New Jersey

Tilbury D (1998) The 'how' of environmental education. Environ Educ, Spring 1998:16–19

UNCED (1992) Agenda 21: the United Nations programme of action from Rio. http://www.un.org/esa/sustdev/documents/agenda21/english/agenda21chapter36.htm. Accessed 10 September 2008

UNESCO (1975) The Belgrade charter. A global framework for environmental education. http://portal.unesco.org/education/en/files/33037/10935069533The_Belgrade_Charter.pdf/The%2BBelgrade%2BCharter.pdf. Accessed 12 September 2008

UNESCO (1978) Intergovernmental conference on environmental education, Tbilisi (USSR) 24–26 October 1977. Final report. http://www.gdrc.org/uem/ee/EE-Tbilisi_1977.pdf. Accessed 12 September 2008

UNESCO (2005) United Nations decade of education for sustainable development (2005–2014): international implementation scheme. http://www.unescobkk.org/fileadmin/user_upload/esd/documents/ESD_IIS.pdf. Accessed 9 September 2008

UNESCO (2007) Highlights on Decade of Education for Sustainable Development progress to date. http://www.gaiaeducation.org/docs/Progress%20to%20Date%20JAN07.pdf. Accessed 12 October 2008

United Nations (1972) Declaration of the United Nations conference on the human environment. http://www.unep.org/Documents.Multilingual/Default.asp?DocumentID=97&ArticleID=1503. Accessed 28 November 2008

Uzzell D, Davallon J, Fontes PJ, Gottesdiener H, Jensen BB, Kofoed J, Uhrenholdt G, Vognsen C (1994) Children as catalysts of environmental change: report of an investigation on environmental education. Final report, European Commission, Brussels

Vare P, Scott W (2007) Learning for a change: exploring the relationship between education and sustainable development. J Educ Sustainable Dev 1(2):191–198

Wals A (1994) Nobody planted it, it just grew! Young adolescents' perceptions and experiences of nature in the context of urban environmental education. Child Environ 11(3):177–193

Wals A (ed) (2007) Social learning towards a sustainable world. Wageningen Academic Publishers, Wageningen

Watts M, Alsop S (1997) A feeling for learning: modelling affective learning in school science. Curriculum J 8(3):351–365

Wenger E (1998) Communities of practice. Cambridge University Press, Cambridge

Whatmore S (1999) Culture-nature. In: Cloke P, Crang P, Goodwin M (eds) Introducing human geographies. Arnold, London

Whatmore S (2003) Hybrid geographies. Sage, London

Williams K, Kalvin E (2007) Relevance in urban environmental education. Paper presented at the Annual Meeting of the North American Association for Environmental Education, Virginia Beach, 13–17 November

Zeidler DL (ed) (2003) The role of moral reasoning on socioscientific issues and discourse in science education. Kluwer, Dordrecht

Zeidler D, Walker K, Ackett W, Simmons M (2002) Tangled up in views: beliefs in the nature of science and responses to socioscientific dilemmas. Sci Educ 86(3):343–367

Index

Lightning Source UK Ltd.
Milton Keynes UK
29 October 2009

145532UK00002B/113/P

9 789048 129553